A QUEST FOR THE PHYSICAL REALITY OF TIME

A QUEST FOR THE PHYSICAL REALITY OF TIME

SIDI BENZAHRA CHERKAWI

iUniverse, Inc.
Bloomington

A Quest for the Physical Reality of Time

Copyright © 2012 by Sidi Benzahra Cherkawi.

All rights reserved. No part of this book may be used or reproduced by any means, graphic, electronic, or mechanical, including photocopying, recording, taping or by any information storage retrieval system without the written permission of the publisher except in the case of brief quotations embodied in critical articles and reviews.

iUniverse books may be ordered through booksellers or by contacting:

iUniverse
1663 Liberty Drive
Bloomington, IN 47403
www.iuniverse.com
1-800-Authors (1-800-288-4677)

Because of the dynamic nature of the Internet, any web addresses or links contained in this book may have changed since publication and may no longer be valid. The views expressed in this work are solely those of the author and do not necessarily reflect the views of the publisher, and the publisher hereby disclaims any responsibility for them.

Any people depicted in stock imagery provided by Thinkstock are models, and such images are being used for illustrative purposes only.
Certain stock imagery © Thinkstock.

ISBN: 978-1-4759-3771-8 (sc)
ISBN: 978-1-4759-3772-5 (ebk)

Printed in the United States of America

iUniverse rev. date: 07/26/2012

ACKNOWLEDGMENTS

The question of time's existence has haunted philosophers and scientists from Plato to Kant. I jointed this club of scientists and philosophers after I received my doctorate in high-energy physics from the University of Minnesota. I started thinking about writing this book in the year of 2001 when I was a visiting-assistant professor of physics at Wittenberg University in the small city of Springfield, Ohio, USA. In fact I came up with the idea when I was writing a paper about time and relativity in the library of Antioch College of Yellow Springs, Ohio. Antioch College is fifteen minutes drive from Wittenberg University. This paper of time and relativity is presented in detail in the appendix of this book.

Sidi Benzahra Cherkawi

First, I want to thank California Polytechnic State University in Pomona, California, for partially funding this book project. Without their help the book would have come out at later time. I want to thank my bright and dedicated student, Jenna Drewitz, for her wise editorial advise, at every stage, on how to improve the manuscript. She helped me edit and check out the manuscript for inconsistencies and redundant ideas. She is studying at California Polytechnic State University in Pomona, California, to become a physician. I want to also thank my previous student of physics, Sofie Leon, who helped me put the first chapters together when I was teaching at California Polytechnic State University in San Luis Obispo, California. Sofie Leon was an excellent student. She is now a graduate research fellow in Civil Engineering, Computational Science, and Engineering Option in the Department of Civil and Environmental Engineering University of Illinois at Urbana-Champaign. I want to thank Thomas Dershem who received his PhD in physics from the University of Michigan and who now teaches physics at the Claremont Colleges in Claremont, California. Thomas and I have shared many ideas in regards to the physicality of time. I want to thank my PhD advisor, Professor Benjamin Bayman, of the University of Minnesota for answering some of the questions I had when I was writing this book. I am grateful to have the chance to share my ideas with my friend, Bayram Tekin who discovered in my paper that space had picked an imaginary part, as he stated it. He is now a professor of physics at Middle East Technical University in Ankara,

Turkey. I want to thank Professor Joseph I. Kapusta of the University of Minnesota for his help in high-energy physics. I came up with my own idea for my PhD thesis but he supervised it to the end with the help of Professor Benjamin Bayman. In facts, Kapusta has helped so many graduate students into completing their doctorate program at the University of Minnesota. Without him and Professor Bayman I wouldn't have gotten my PhD. I want to thank Nobel Laureate, Dutch physicist, Gerard T Hooft for sending me some problems to work on. I want to thank Nobel laureates, Leon Lederman, Murray Gell-Mann, T.D. Lee, who answered some of my physics questions during my stay at Columbia University. I want to thank Nobel Laureate, I.I. Rabi (deceased) who gave me an advice, which often helped me through out my academic career, and which I had discussed it in this book. And finally I want to thank Nobel Laureate, James Rainwater (deceased) who told me to use the operator THINK before I ask any question in physics and who gave me many ideas in physics that I am still using to this day. I did very well in his physics class and later on when I moved to the University of Minnesota to pursue my graduate studies and I needed a letter of recommendation from him, his secretary told me that he had died. The operator THINK helped me write this book. Finally, I want to thank my friend, Abdelhaq Boudlali, for allowing me to use him as an ear that listens to me whenever I want to discuss the subjects in this book.

"Yet as Einstein and Gödel well knew, it is not space but time that in the end poses the greatest challenge to science."
Palle Yourgrau (A World Without Time.)

Dr. Sidi Benzahra Cherkawi
A Quest for the Physical Reality of Time

CHAPTER 1

The Hungry Mind

As a physicist, I am always trying to stay practical. Physics is a very practical subject, and that's what makes it very useful. Without physics there will be no plane to fly or phone to call your friend or computer to surf the web. If a scientist starts to philosophize, often times, an experiment will be conducted and it will refute that philosophy. Several physicists I encountered in my life have died, and unfortunately cannot defend themselves if I say something untrue about them in this book. But, I have held my opinion amounting to conviction that if I lie or cheat, my mind will be filthy, and knowledge or wisdom, which are clean and pure, will not come to my mind. So I am sure that everything I wrote in this book is true to the best of my knowledge and memory. Any falsification was not intentional. I read several books and articles when I was thinking about writing this book of time. So some ideas in this book are mine but some belong to other people. Like John Donne, the famous poet, said, "No Man is an Island." Our ideas are built on other ideas. Some are ours, some are ours mixed with theirs, and some are theirs. This book is not written to make me money or bring me fame. I care less for money and fame. This book is written to help the coming generation build on it, to help humanity understand this world. The more we understand this world the more we make life easy for us.

In the early eighties when I was attending Columbia University as an undergraduate student in New York City, I was rich in spirit. There was a fire burning in my chest, which still is to some extent. I wanted to attain all the knowledge in the world. I wanted to sit down and have a cup of coffee with God and ask him questions about the big bang, about where I came from, and the purpose of my life. I wanted to know how I got here and where I was headed. I wanted to know if there would ever be a scientific possibility for me to live forever—die by accident only. I wanted to know why I was a human being and not a bird or a cat. I wanted to know all the answers to the important questions. I wanted to know everything. That's how much fire I had in my chest. This fire was burning inside me with a relative infinite-source of energy. Even though I am getting older, I still have this fire burning in me now as I write this book. I want to know why plants are green and why the color green. I want to know space and what is made of. I want to know if time is physical, such as matter, or a figment of our imagination such as the numbers 1, 2, 3. I figured by writing a book I would learn and discover more as I read, think, and dig inside my mind and the minds of other people. I couldn't wait to learn.

Columbia was, and is, an excellent school. It was surrounded by big sturdy buildings—still is—as though to isolate its students from the outside world. It had a lot to offer when I was attending. I remember one time, I asked the librarian about the number of books I could check out, and she said "as many as I wanted." So, I checked out a bunch of books and sat there on the steps in the mall reading one with a stack of the others on my knees. As a kid growing up in Rabat, Morocco, I used to go to this library downtown where I was allowed to check out only one or two books at a time. I remember the books were covered with transparent plastic and filled with beautiful pictures. I loved those books. So, this idea of unlimited books at one time was such a shocker to me. When the librarian told me this, I seized the moment every weekend and checked out so many books to the point, that while I carried them home they were stacked to my chin. I looked at the stack of books and felt very emotional. I didn't know why I felt so emotional. A clinical psychologist could probably figure that out. My explanation is that I felt like a kid who wanted a toy that he couldn't have. There was all that knowledge in those books and I was sad that I couldn't have it in my head. I knew that I wouldn't know it just by reading it. You need to have a sharp mind to understand it. Physics is not like a novel. You can't just read it and know the story. I read many novels. Reading certain novels is like drinking water—simple. Learning physics is like trying to know the mind of God, trying to find out what is inside the mind of the creator or Nature or whatever you believe is divine in your scripture. I was trying to learn what was inside the mind of something very challenging. You have to keep asking, thinking, and doing experiments to see if that thing is telling you the truth. It is a long, tedious process.

The knowledge was in the books, the books were on my lap, yet the knowledge wasn't in my head. I knew it would be a long time before I would possess this knowledge. It is like we have this talisman in the palm of our hands, but it takes us years and years to decipher its meaning in order to understand it completely. I attended an elite university to increase my probability of meeting and interacting with great people. Of course there are great minds that have never gone to school. There were minds before there were schools.

In my days at Columbia University and the University of Minnesota, I encountered great physicists. There were I.I.Rabi, Leon Lederman, T.D. Lee, James Rainwater, Stephen Weinberg, Murray Gell-Mann, Norman Christ, Joseph Kapusta, Benjamin Bayman and more. I will talk about them and share some of the things that I have learned from them throughout the years.

I always talked to Isaac Rabi at Columbia University. He was a friend of Einstein. There were other good scientists at Columbia University at that time, but I preferred Rabi because he was old and wise and easily accessible. Old people care less about their time. Some people dislike old people, but they should realize that old people are either smart or lucky, two reasons I believe they live longer. If they had been stupid they would have died earlier. I always considered old people to be wiser. My neighbor of many years, May Clutter, was old and wise. She always knew what plants would survive in her garden and what plants would die. She always knew which direction storms were coming from and where they were headed. One day when there was a storm, I climbed a ladder to clean the gutter on my house when May reminded me that I could get hit by lightning. I knew the physics of lightning and how it works, but I forgot about its effect, something May didn't.

I.I. Rabi did a lot of research in physics. Some of his research was about magnetic resonance imaging, a technique that is widely used in hospitals today. A few months before I.I. Rabi died in early 1988, his medical doctors put him inside a magnetic-resonance-imaging machine. Once inside the machine, Rabi saw a distorted image of his face in the curved wall surrounding him. "It was eerie," Rabi said. "I saw myself in that machine. I never thought my work would come to this."

I.I. Rabi had done great work in physics and he was, and still is, famous in the physics community. I remember once, Stephen Weinberg, from the Weinberg-Salam Model, came with his wife to give a speech at Columbia University. I can recall his wife wearing a fur coat. The speech was held in the evening in the mall of the university near the Pepin physics building and there were many people sitting in their chairs, listening to him. He said that he needed to watch carefully what he was saying because I.I. Rabi was sitting in the audience.

I.I. Rabi had a large office in the Pepin physics building; he had a large desk and a large leather couch. The couch was dark brown. Most of the time he had the curtains drawn down, which made the inside look dark and feel cool. I would accompany him whenever I saw him walking to the physics department. He walked slowly because of age and when he talked he moved his lips as though something was stuck between them and he was trying to get it out. He always wore this hat that city people in America use to wear in the forties and the fifties. I once took four physics classes and didn't do as well as I was expecting in them. I approached him and asked him why he thought I didn't do as well, and he reminded me that I had taken four hard classes. He said that I should have taken some art classes, or literature along with physics. When you go for lunch, he said, you don't just eat meat or the main dish. You also eat salad and dessert. Registering for classes, he said, is like going for lunch. He also said that if you want to be successful in your career you need to do what you are good at and not always what you like. Just because you like something, doesn't mean that you are good at it. Later on I was contacted by the school administration and the clerk told me that I had to take some elective classes such as music or art. I remember she told me, "you can't just use the right hand; you need to use the left hand too." I took music where I learned about the Gregorian chant, something that never helped me in my scientific career, and took an art class. The art class was fascinating. We used to meet every Saturday morning in the Geology building—I don't know why my art class was in the geology building—and sit around a big circular table with other students and watch slides of Roman and Greek buildings and sculptures of famous Italian Artists while eating a poppy-seed muffin and drinking coffee from a paper cup. That was my routine every Saturday morning and I loved it. In facts, I enjoyed that class more than any other class through out my academic career.

I.I. Rabi prepared me for my academic career. When you are good at something you do well in it. It is true. One day I sat down and tried to find out what I was good at. I came to the realization that I am good at one thing, helping people understand things. I cannot say I am good at physics because there are so many physics problems in my head right now that I cannot solve. In facts, I was given some high energy physics problems by the Dutch Nobel laureate Gerard 'T Hooft to work on, but I decided to pursue the problem of time instead. I.I. Rabi's concept of "don't do things you like that you are not good at" has stuck with me and helped me through out my life. I have liked so many things that I could have pursued as a career, but I wasn't good at them. I.I. Rabi gave me a path to follow in life and this path has led me to write this book and to have a peaceful healthy life, so far.

CHAPTER 2

The Absolute Frame of Thoughts

In physics, we always have to have a frame of reference to help us observe and measure things. Without a reference frame you cannot tell if something is big or small. You cannot tell if something is moving or standing still. Imagine yourself in space alone with no starts around. You will not be able to tell if you are moving or staying still. Perhaps, if you accelerate you might feel a jerk in your guts but if you are moving with constant speed you will not be able to tell if you really moving. For example, while watching the sunset in the horizon you can detect the motion of the sun—which is actually the motion of Earth—because of the line that forms the horizon. If you take off that line you won't be able to see that motion. The horizon here plays the role of a line of reference. If you are driving a car at 70 miles per hour and somebody who is moving with respect to you measures your speed in a moving frame, he or she will find a different speed. The speed will be either greater or smaller, depending on the direction of the moving frame. Now you ask the question, who is right, you or the person who is moving? The truth is, they are both right in their own frame. Now you ask yourself can this phenomenon happen to a thought? For example, when you think of something, you are using a frame "located" in your brain that helps you weigh in your thought. This frame helps you measure the validity or the strength of your idea. Sometimes you come up with good ideas and other times, bad ones. Now when you share your idea with somebody else, that person will also assess your idea, but he or she will assess it using his or her own frame of reference, or her own frame of thoughts. That person might not agree with you. You might say your idea is good, but he or she might have a different opinion. Now who is right? In your own frames, you both think you are right. This can create discontent because you want to know who is truly right. If you kill somebody in Texas, the state will kill you, but if you kill somebody in Minnesota, that state will confine you until you die. Which state is right? Both states are right in their own frame of thoughts. Here is another controversial example about the frame of thoughts: If you have sex with a fifteen-year old in California, the state will put you in jail—A film director by the name, Roman Polanski, is banned from entering the United States for having sex with a minor in California. My mom, on the other hand, who is Moroccan, had her first child when she was 15. She has nine kids total. And I am her third child chronologically and she is only nineteen years older than I am. My mom is still alive and we go for walks all the time. Who is right, Morocco or California?

To find the true answer you need to have an absolute frame that can measure your thought. The problem here is there is no such frame. This frame doesn't exist. So we will never know the true value of a thought. We can only know its strength relative to another thought. What we know is the comparison. We can compare two thoughts, but we cannot truly evaluate one single thought. This is the same as if I were to show you a table by itself in space. You won't know if it is big or small. But, if I brought another table next to it, you would be able to compare the two tables. This is the problem I have been encountering in my academic career, let alone my life. I want to have an absolute frame of thought. I want to know the exact value of a thing. I realized that whatever I think of, is not truly known or absolute. This book I am writing now is filled with ideas that are not absolute, ideas that are measured in a frame with unknown whereabouts. The fact that I can compare things gives me hope, courage and persistence to keep investigating this problem of time. I can compare the values of two observables, such as the length of two tables or the mass of two objects. We don't know how big is "big" and how small is "small", but we can compare the big to the small. All what I was writing here was set in stone by Werner Heisenberg in his Uncertainty principle which states that one cannot know the exact value of things, and also set by Kurt Gödel in his incompleteness Theorem, which states that our knowledge of the basic truths of mathematics is limited. Here we get two constraints: one physical and the other is mathematical.

In regards to time, I developed a thought experiment that will assist me in comparing the physicality and non-physicality of time. This thought experiment will change into a physical experiment, which I will conduct in the future. Now, if we find that time is not physical, we will have to tweak our previous physics theories in such a way as to make time irrelevant. G.T. Whitrow stated, in his book, The Nature of Time, on page 165, that Parmenides concluded that time does not pertain to anything that is truly 'real', but only to the logically unsatisfactory world of appearance revealed to us by the senses.

Parmenides' belief that temporal flux is not an intrinsic feature of the ultimate nature of things has been tremendously influential. It is not only idealist philosophers who have claimed that the temporal mode of our perception has no ultimate significance. Even so empirically-minded a thinker as Bertrand Russell made the following admission in his well-known essay on 'Mysticism and Logic': 'There is some sense-easier to feel than to state-in which time is unimportant and superficial characteristic of reality. Past and future must be acknowledged to be as real as the present, and certain emancipation from slavery to time is essential to philosophic thought. Herman Weyl seemed to agree with Bertrand Russel. He wrote that the passage of time is a feature of consciousness that has no objective counterpart.

The emancipation from slavery to time is not only essential to philosophy, but also to physics. Now how can we do away with time in physics? Doing away with time in physics is like doing away with money or credits in economy. You can tell if somebody is running faster than somebody else without using time. The person who runs fast arrives at the destination first. If you observe them, you will notice that their legs and arms are moving faster. If we can tell that one person is moving faster than the other without using time, we can conclude that speed can be described in another form or a formula that will not include time. Now somebody might ask us the question: how fast is the faster person moving? This can be a difficult question to answer if we don't include time. So, to solve this problem, we can develop a unit of speed and write each speed in terms of this unit. Physicists have already implemented this idea. For example, the mach is a unit that describes the speed of sound. One mach is about 330 meter per second, depending on the temperature. The speed of sound depends on the density of air. It also depends on the elasticity or, to be scientifically correct, the bulk modulus. The speed of sound in water is larger than the speed of sound in air. Density and elasticity both compete to change the speed. The density tries to decrease the speed and the elasticity tries to increase it.

The speed of sound is faster in water than in air because of the density and the elasticity. The speed of sound in water does not depend on time. It is a physical process, or an interaction like Boltzmann had stated. In physics there is a formula that ties speed to time. There is also a formula that ties speed to acceleration without the use of time, but one needs to know that the acceleration itself depends on time. It can be described as the change of speed with respect to the change in time. Another way of avoiding the use of time is by picking a certain speed that we are aware of in nature and using it as a reference or a factor, like we do with the mach, in order to jot down any speed we might find in terms of this chosen known constant speed. Any speed in the universe can be written in terms of this speed. For example, we can use the speed of Earth with respect to the Sun and call that speed "Earth Speed with Respect to the Sun" and give it the acronym, ESRS. We can also use the speed of the electron of a stable hydrogen atom as a reference, like with ESRS and mach. Now all the speeds we know can be written in terms of the speed of the electron. The speed of light can be written in terms of one of these two speeds I talked about. So far, I didn't do anything useful in finding the speed without the use of time. All I have done is state that any speed can be written in terms of the speed of Earth around the sun, or in terms of the speed of sound or the speed of the electron going around the proton in the hydrogen atom. Now all we have to do is find a way to dissociate Earth speed from time. If we find that Earth speed is dissociated from time or doesn't need time at all, we will be forced to believe that all other objects' speeds can also be dissociated from time. We can state that when Earth goes around the sun, it cuts into space. Space is the only dimension involved in this process. Of course, somebody might ask the question, how fast does Earth cut into space?

CHAPTER 3

Does the Electron Get Old?

Descartes never liked the idea that people think about infinities. Since we are finite, he said, it would be meaningless for us to verify anything concerning the infinite. It turns out that the human mind, though limited, can consider the infinite. For example, you don't have to literally go to the sun and stick a thermometer inside it to know its surface temperature. We can calculate the surface temperature of the sun on the back of an envelope. In facts I always give this problem to my physics 132 students at California Polytechnic State University, Pomona. By knowing the temperature of the surface of the Earth and the distance between earth and the sun, we can determine the temperature of the surface of the sun, which is about 6,000 degrees Kelvin. Since the physics is connected in a way such that we can know the temperature of the surface of the sun, without us going near the sun to measure it, the mathematics can possibly be connected in the same way such we can measure infinities without us having an infinite number or a thing in our disposition. We know that when we divide one by an infinite number we get zero. It is like slicing a pie into an infinite number of equal pieces. There will be nothing left to eat in the end. We can have an idea of infinity even though our mind is not infinite and does not contain infinite knowledge. For the same reason, like Descartes thought, people might not like the idea that I question the existence of time; for this could be seen as an attempt to define it in order to deny its existence. But does time really exist physically as does matter?

Of course we know that time exists in some sense. We also know that numbers exist, but numbers exist only in our mind. Take pi for example. During the dinosaur era there was no Pi. My question is, does time exist the same way matter exists or the same way numbers exist? There is a school of thoughts that states that time is physical and it exists the same way a person exists or matter exists. There is another school of thoughts that states that time was invented the same way money and the numbers 1, 2, 3 were invented. Money and numbers didn't exist before humans came to Earth. Matter was here before humans, but was time here before humans also? This is what I mean when I ask the question: does time exist?

Physics without an experiment becomes philosophy. To come up with an experiment that could prove or disprove the existence of time, we first need to know if time has an effect on physical things. Does the passing of time make people old, or do people get old regardless of time? Do animals migrate under time schedule or do they migrate because of the change in the environment around them? Do plants germinate under some time plan or do they germinate under the environmental situations they are in? I asked a friend of mine, Abdelhaq Boudlali, who is a biologist and studies plants at Minnesota Agriculture Department about this and he told me that plants go with the feel and not with time. For example, if you fool a plant into believing that the spring is coming, the plant will bud. A stretch of sun can bring plants to bud fast, but cloudy days delay bloom. This shows that plants don't use our time to keep track of their growth.

Does time have an effect on people? People who don't take care of themselves look older than people who do take care of themselves. For example, people who do hard labor shorten their life span by a considerable amount. Consider mining where workers have to breathe in all that soot and metal particles and work in various temperatures. People who expose themselves to the sun get more wrinkles than people who wear protective clothing, use sunscreen, or stay indoors. I once saw a picture of the face of a US trucker whom the left side of his face shows more wrinkles that the right side. And the reason was because here in the US, we drive on the right side of the road so the driver sits on the left side of the car and his or her left side gets more exposure to the sun than the right side. Also The amount of damage to the skin caused by the sun is determined by the amount of radiation exposure and the person's pigment protection. In this case, we can state that the amount of radiation from the sun had an effect on aging. But is time relevant here? Some people might say that the person who works out in the field received too many wrinkles because he or she stayed exposed to the sun for a long time. Another person might say, no, it is because a large amount of radiation has been absorbed by his/her body. You get more wrinkles by staying exposed in the hot sun for a short time than by staying in the soft sun for a long time. The amount of radiations received per unit area of the skin is the scenario that should be taken into consideration. So time here is irrelevant. Here is another example, if you constantly switch the light on and off in a light bulb, the light bulb will not survive that long. The constant switching of the light perturbs the equilibrium of the light-bulb-room system. This lack of equilibrium interferes with the lifespan of the light bulb. In the same fashion, highway mileage in a car contributes less stress in the engine, as does street mileage. When you drive in the highway, you tend to stay at constant speed and this causes equilibrium in the engine. The pressure in the engine and the flow of gasoline are occurring at constant rate. On the other hand, stop and go street mileage causes to accelerate and break often and this leads to more stress on the engine, which in turns causes instability and lack of equilibrium. So the engine of a car ages because of the wear and tear not because of time. If you buy a new car and you store in vacuum where there is no air to interact with and you cover it so that that radiation won't get to it, and isolate it from chemical reactions that can occur in it, the car will not age.

If time is not relevant to our aging, then could our existence be timeless? The physicist Erwin Schrödinger once stated that Plato is renowned even though he made no special discovery about numbers, formulas, or geometrical figures. What made him famous, Schrödinger said, was the fact that Plato was the first to imagine the idea of timeless existence and to state its reality as more real than our actual experience. One can ask, what is an example of something that is timeless? One thing that comes to my mind is a mathematical truth. A mathematical truth is timeless; it does not come into being when we discover it. Yet its discovery is a very real event. One might argue that real numbers have no physical existence. They are a human creation and thus the product of a biological and cultural creation. So, a mathematical truth that involves real numbers can be itself a human creation. Pi is a real number. Is Pi a human creation? The Babylonians and the Ancient Egyptians discovered Pi a long time ago. Pi is a mathematical truth. This mathematical truth is always true. It doesn't need time to exist. It doesn't need anything to exist. It doesn't get old. It is timeless.

I once attended a seminar in the Physics Department of North Dakota State University in Fargo where a colleague from the Chemistry Department made a presentation about corrosion, and how humidity and heat affect the paint of a car. He showed us pictures of a car and a military aircraft with corroded paint and spoke about how cars and planes could easily get corroded. I approached him after his speech and asked him, "If I put a piece of metal in a vacuum, in empty space, will it get old?"

He looked at me and said, "Get old or get corroded, which one do you mean?"

I said, "Get old."

He thought for a moment and replied, "I don't know."

If, for example, we take a piece of wire and put it in a pure vacuum, will it get old? If it gets old, how does it happen? A piece of metal is a collection of protons, neutrons and electrons. Will the protons, neutrons and electrons get old? In general, does an electron get old? Physics tells us that an electron that was made billions of years ago right after the Big Bang is the same as an electron that is made today through an interaction. The electron has an intrinsic energy that is always constant regardless of time. It is like a person who never gains or loses weight. The electron also has an intrinsic spin. This doesn't mean that it is physically spinning like a wheel; it is just a characteristic that Wolfgang Pauli came up with to differentiate electrons. He discovered that two electrons cannot be in one state—a mode or condition of being—or cannot go through each other or merge together as two ghosts, so to speak, and therefore there has to be something different about the two electrons for them to be close to each other and in the same state. One electron will spin up and the other will spin down. Because they are not physically spinning, we can imagine one will spin clockwise and the other counterclockwise.

This brings me to the time when I was an undergraduate student at Columbia University. Dr. Chien-Shiung Wu was a Chinese American physicist who conducted an experiment for Tsung-Dao Lee and Chen Ning Yang, two other Chinese American Physicists. They showed that parity is not a conserved property and received the Nobel Prize, but sadly, Dr. Wu didn't receive it. I approached Dr. Wu as she was leaving the physics building. I asked her, "Why do electrons have spin?" She stood with me outside for a moment, thinking. I remember the sky over Columbia University was blue with a few white clouds overhead. She looked at the white clouds and said, "It is like the clouds, they move," and she gestured with her hand and forefinger, making a circular motion, indicating the spinning. That's all I could get from Dr. Wu, which disappointed me because I wanted to know where the spin came from. I didn't want to keep on asking her because she was tired and withdrawn. Now, the reason I am hung up on the spin is because the strength of the electron's spin doesn't change. It is independent of time. Physics tells us that the strength or the magnitude of yesterday's electron spin is the same as today's. So why doesn't time affect the spin? We are made of electrons. We grow old and die but the electrons in us do not grow old and die. We are made of something that doesn't die but we die. So do we grow old because of time or because of the change of the mechanism that runs our body? We say that we grow old because of time but how come some particles don't grow old. Gold doesn't grow old. Gold is always gold. In fact, some astronomers have recently suggested that gold was probably made in collision of neutron stars. They have stated that Earth does not have enough pressure to create such a neutron-rich element as gold. It is possible that you already own a souvenir from one of the most powerful explosions in the universe. Metals don't get old; they get corroded. If, like I said before, I put a piece of metal in a vacuum away from radiation and humidity and other corrosive materials, the metal will not get corroded or grow old. It will stay the same. Where is the effect of time here? Electrons do not decay. So, if electrons do not decay, they are not affected by time. Time is changing but the electron is not.

Of course, right after the big bang there was no electron. There was just a cloud of quarks and gluons, called the Quark Gluon Plasma. The Quark gluon plasma is a hot gas of quarks and gluons. It is plasma because it conducts currents. Physicists are now trying to make one temperature of 1 GeV, which is approximately 1 billion times hotter than the surface of the sun. The temperature of the surface of the sun is about 6000 Kelvin and that of the interior is about 15000000 Kelvin. Quarks have never been found free by themselves in ordinary matter, for they are always held together by gluons. But by smashing heavy nuclei of gold or lead, each ion coming at each other with 7 TeV energy, we expect the nuclei to dissolve into hot plasma of free quarks and gluons. This is exciting because we have already heard that a quark cannot be isolated.

We can ask ourselves, why is it important to study quark-gluon plasma and the physics related to it? The answer is that we need to understand the conditions the universe was in when it was only one millionth of a second old. This might shed some light on where we came from or how we came about. Moreover, the study of quark-gluon plasma will shed new light on the theory of quarks and gluons, quantum chromodynamics, known as QCD. QCD is a precise, quantitative theory, but many of its details are still misunderstood. It is more complicated than QED, quantum electrodynamics, which was carefully studied by physicists such as Enrico Fermi and Richard Feynman.

The minimum temperature required to achieve de-confinement of quarks is generally understood to be about 150 to 200 MeV. RHIC, the Relativistic Heavy Ion Collider at Brookhaven National Laboratory on Long Island, New York, is the first collider designed to specifically create this plasma. It may reach a temperature of 500 MeV. CERN, on the other hand, hopes to reach a temperature of 1 GeV by colliding heavy nuclei in the Large Hadron Collider. The collisions will be studied with advanced detectors for high-energy particles. These devices have been designed and built by scientific collaborations involving nearly 100 universities and laboratories worldwide. Continuing our discussion about the electron, it has an intrinsic energy that is always constant regardless of time. The electron also has an intrinsic spin that is related to this energy.

Was time created with the quark gluon plasma? If it was created with it, how was it tied to it? What was the relation between time and energy back then?

CHAPTER 4

My Meeting with Nobel Laureate Murray Gell-Mann

We need to ask ourselves where the idea of time came from. In 1995, I met Murray Gell-Mann, the father of the quark model and Nobel laureate, at a talk held by the University of Minnesota Physics Department. I was a graduate student at that time. He came to the Theoretical Physics Institute to promote his book, The Quark and the Jaguar. First, I wanted to challenge him by asking him a philosophical question: how do we know that what we think of is what there is? He understood my question and quickly answered "we will never know." Secondly, I asked him an innocent physics question: Where did the Big Bang start? He answered me by saying, "You cannot ask that question because there was no space before the Big Bang, which means that there was no "where"." He also said that there was no time before the Big Bang, which means there was no "when". If time did not exist before the Big Bang, was it created during the big bang?

If time was created, we need to know from what it was created. There are two fundamental things in the universe: space and energy. Some philosophers argue that space doesn't exist, but that should be another topic of discussion. I wish somebody would write a book titled, A Quest for the Physical Reality of Space. Was time created from space and energy or was it created from something else that we don't know. Logic tells us that things get created from other things. One cannot create something from nothing. Physics is against the assumption of things getting created from nothing. The Second law of Thermodynamics states that nothing can get created or generated without a loss of energy. You cannot make things move without supplying some amount of energy to them. Whatever created time needed energy to create it; but before the big bang there was no energy. Of course there are physicists who will tell you that there was some energy in nothingness before the big bang started. This will conflict with what Murray Gell-Mann said and what many physicists had said before him. Try to think of something that is created from nothing and you will find out that there is no such thing. If you study all things that exist in the world, you will find out that some energy had to be spent for that thing to get created. Even an idea that is not physical has to come out of some process that needed an energy of some sort. Again, we can ask ourselves this question? Does time have a physical property, like matter? Why is it important to find out whether time has a physical property? If time has a physical property, it might interact with matter. For example, we know that matter is physical. If you see a car coming at you, you move away because if you don't, the matter of that car will interact with the matter of your body. If time has a physical property and interacts with physical objects, its interaction will be real and the consequence of that interaction will also be real. Time, in this case, might interact with matter and space and our understanding of these interactions will be based on something that really exists physically and not on something that is the product of our imagination.

Usually, things that exist in a physical world will interact with that world. For example, Dark Matter or Dark Energy interacts with our world by making our universe expand faster, even though we cannot see or detect this energy. It exists because we can see its effects. Where is the effect of time? If time physically exists in this world, it would interact with this world like Dark Energy does or in another form. If time really does interact with this world, it does so differently than other physical entities, such as matter. Take the idea of time reversal. Think of it as if you run a movie backward; nothing physical happens to the actual film. Physicists agree that time reversal does not affect any physical process. In all the laws of physics that we have found so far, there doesn't seem to be any distinction between the past and the future. The laws of physics still hold even if a physical process goes backward in time. There is a one-to-one relationship between a process and its time reversal. If a dish falls down from a table, it can go back up onto the table as long as you give back the energy lost to the act of falling. But if you take away some of the energy from the dish-table system, the physical laws of Nature will not allow the dish to return to the table and sit on its surface the same way it was sitting before the act of falling. Physicists assigned this time reversal phenomenon the letter "T". It is like an operator that acts on a process and changes it; it will make time run backward. If this "T" acts on a car moving forward, the car will go backward. Some physicists might bring the idea of entropy and state that some processes are irreversible and cannot go back to their original state, but if you allow them more time and give them back their lost energy, these processes will go back to their initial states.

Now we can return to the question: How can we prove that time is physical if it is physical? In general, if a scientist wants to prove that something is physical, he or she needs to conduct an experiment or write a theory that will be tested by an experiment. But the first step is to consider probabilities. This is not a simple situation we are in. For example, if you want to know if a table exists in a room, you would enter the room and touch that table. After you touch the table you know that it exists in the room. The fact that you touched the table generated an interaction between the particles in your hand and the particles in the table and that interaction gave you the feel of the touch, which means that the interaction made you sure that the table exists. In our situation we are dealing with time that we cannot just touch to determine its existence. So, we have to come up with a different experiment to prove its existence. To prove that an atomic particle exists, for example, a scientist needs to first calculate the probability for such a thing to exist. Unfortunately, dealing with mathematics, almost anything has a probability of existing. And on the contrary, Kurt Gödel has already proved to us in his incompleteness theorem that there is always some limit in our logic when we try to prove something mathematically, so it is possible that our mathematical proof of the physicality of time will fall within an interval of uncertainty in logic.

You might not believe this, but there is a probability that a bunch of monkeys can write the novel, Grapes of Wrath. It was the novel that led John Steinbeck to win the Nobel Prize in literature.—I read almost all of John Steinbeck's books. I wasn't aware of who John Steinbeck was. I just read his books one after the other. I would buy each copy of his books for 25 cents from a Goodwill store in Minneapolis. I found his books to be easy to comprehend. He writes in simple English and since English was not my first language, I stuck to his books. I didn't know he was that famous until I met a girl at school who taught me all about him. I used to feel the same about Stephen King, but when I read his books I noticed something very interesting. I noticed that Stephen King had been reading John Steinbeck's books and they influenced him. I could tell from the expressions Stephen King wrote.

Going back to the idea that a bunch of monkeys can write the Grapes of Wrath, if we give ten computers to ten monkeys, and show them how to type on the keyboard without teaching them English or any other language, the monkeys will type the Grapes of Wrath; but it will take them a long time. How will they do it? The Grape of Wrath starts with this statement: "To the red country and part of the gray country of Oklahoma, the last rains came gently, and they didn't cut the scarred earth." As they start to type, there is a probability that the first monkey will type the letter in the book, which is T. There is also a probability that the second monkey will type the second letter in the book, which is o. And there is a probability that the third Monkey will type the third letter in the book, which is t, and so on until we get to the tenth Monkey. There is a probability that the tenth monkey will type the letter o that is in the word country. After the tenth Monkey is done, we go back to the first. There is a probability that he would type the letter u to start completing the word country. But the catch is, they have to keep on typing letters for millions and millions of years. The important thing, however, is that there is a probability for the Great American Novel to be written by a bunch of monkeys. This sounds strange, but mathematically there is a probability that it can happen. As I said before, if we want to prove that a particle exists in an atom we must calculate the probability for it to exist, and that is exactly what we do now in quantum mechanics. We only calculate probabilities.

CHAPTER 5

My Meeting with Nobel Laureate T.D. Lee

When I was an undergraduate student at Columbia University, I was bothered by this idea of probability. The thing that bothered me most was the idea of cube probability. For example, if you put 10 white cubes in a box, each cube having a number of 0-9 written on it, and you calculate the probability of pulling the cube with number 2, for example, you will find the probability to be 1/10 even though it is possible that you will pick number 2 cube in the first shot or the 100th shot. The probability is fixed even though the number of trials in picking the number 2 cube can be different.

At the time, there was the famous Chinese physicist, T. D. Lee, working in the physics department. I once saw him walking up the stairs and asked him,

"Why is it when we want to calculate the probability of some tiny physical process, we take two wave functions—a wave function is a mathematical task or utility that has information about an object or a particle—we multiply them and we sum the results of the multiplication over some volume?" He responded by saying that "physics is inductive." And walked away. He didn't even give me a chance to react. Even though I had taken some philosophy in high school, I didn't know much about inductive and deductive reasoning. This is the subject you would know about if you studied philosophy. I thought about his answer and came up with the conclusion that if, for example, you state that all men have two arms and John is a man, you can deduce that John has two arms. This is called deductive reasoning, which is different from inductive reasoning where you check each man and determine that many of them have two arms and then declare that John has two arms since he is a man himself. Both reasoning are often used in physics and mathematics. This is how we conclude that copper is conductive, or conduct electricity. We take one piece of copper and insert it in an electric circuit and find out that it conducts electricity and then we take another piece of copper and do the same to it and we also find that it conducts electricity and we do it for the third piece and so on. We can then conclude that all copper pieces conduct electricity. This is called inductive reasoning. It works because there is no piece of copper in the world that won't conduct electricity. This is what T.D. Lee was talking about when he stated that the probability of finding a particle in some location is derived from some inductive reasoning. Quantum mechanics can be at least partially based on inductive reasoning.

Quantum mechanics does not tell us where a particle is exactly, it only tells us a region where it can be. So, we might state that if we want to know if time physically exists we need to calculate the probability for it to physically exist. Finding the probability is good, but not good enough.

To prove that time exists or doesn't exist physically we need to conduct an experiment. We can do a theoretical calculation, but if an experiment doesn't support it, the theory merely becomes an idea or a model. It doesn't mean that the theory is wrong, or the idea is wrong; it just means that the theory cannot be fully trusted. Scientists usually do not trust a theory that is not supported by an experiment. Now, there is this theory of Super-symmetry that scientists are talking about. Many physicists are working on this super-symmetry, but many other physicists doubt it because it is not supported by an experiment yet. The Super-symmetry physicists state that there are these super-symmetrical particles in the universe. As of now, the experimentalists couldn't detect one yet. If no super-symmetrical particle is detected, the super-symmetry will become just a beautiful mathematical approach to the understanding of the universe.

Scientists used to think that light needed a medium to propagate in space, the same way sound needs air to propagate from a musical instrument to your ear. When you say the word "go" for example, your vocal cords exert a small pressure on air and that pressure propagates in the form of a longitudinal wave—if you pinch a spring you will see that the squeeze that you caused to the spring will propagate along that spring. This is called a longitudinal wave. This sound wave propagates through the air in somewhat a similar fashion to water waves, which are actually transversal—When you drop a stone onto a water surface the water goes up and down as the wave propagates forward. If there is no air, the sound will not be able to travel, and therefore it cannot be heard regardless of how loud you yell. Because of this, and probably other reasons, scientists believed that light might also need a medium, called ether, to travel. Light propagation turned out to be quite different than sound propagation. Light is a photon or a bundle of energy, so to speak, that travels in space in the form of two fields: one electric and one magnetic. In some orientation in space, the electric field oscillates vertically and the magnetic field oscillates horizontally. They both oscillate as light propagates through space. The inclusion of ether was a model. Scientists needed to conduct an experiment to prove whether light needs ether to propagate in space. Two scientists, Albert Michelson and Edward Morley, proved that light doesn't need ether to propagate in space. Light can travel in a vacuum, void of all particles. We need to conduct an experience, as did Michelson and Morley, in order to prove or disapprove the physical existence of time.

As stated in Schrödinger's book, What is Life?, Ludwig Boltzmann believed that time has no direction; time is just a group of statistical considerations. If we take a deck of cards that is ordered (i.e. 1, 2, 3, 4, 5, 6, 7, 8, 9, 10) and we shuffle it once or more, the order will turn into some random set (e.g. 2, 3, 8, 6, 1, 7, 5, 9, 4, 10). This is not an intrinsic property of the process of shuffling. If we get a set of cards that is disordered, a process of shuffling can cancel the effect of the first shuffling and restore the original order. In our example, the arrow of time is not worked into the mechanisms of interaction represented by the mechanical act of shuffling. This mechanism does not make use of the notion of past and future; it is in itself completely reversible. The arrow, the very notion of past and future, results from statistical considerations. In our example with the cards there is only one, or very few, well-ordered arrangements of the cards, but billions and billions of disorderly ones. If we continue to shuffle the cards, we can get back to our one to ten ordered set. Some clever people opposed this idea and wondered: If the shuffling of cards is symmetrical, why is the whole system going in one direction? Boltzmann found the answer; we need more time. If we have enough time we can get back to the set we started with. Boltzmann maintained that if the universe is sufficiently extended and exists for a sufficiently long period, time might actually run in the opposite direction in distant parts of the world. Schrödinger agreed with Boltzmann by stating that on a very small scale, both in space and in time, such reversions have been observed in Brownian movement.

This is not a new idea, as stated in G.J. Whitrow's book, The Nature of Time. According to Nemesius, the Stoics believed that Socrates, Plato and each individual man will live again with the same friends and fellow-citizens. They will go through the same experiences and the same activities. Every city, village, and field will be restored, just as it was. And this restoration of the universe takes place not once, but over and over again, indeed to all eternity without end. The Stoics believed that those of the gods who are not subject to destruction, having observed the course of one period, know from this everything which is going to happen in all subsequent periods. There will never be any new thing other than that which has been before, down to the minutest detail.

CHAPTER 6

You Need to Rape that Paper

In his theory of relativity, Einstein took time as a given and worked with it in his thought experiments. He found out that time can be written in terms of a unit-less expression called the Lorentz factor and denoted by the letter gamma. The Lorentz factor or Lorentz term appears in several equations in special relativity. It comes from deriving the Lorentz transformations. The name originates from its earlier appearance in Lorentzian electrodynamics—named after the physicist Hendrik Lorentz. If the velocity of an object changes, then gamma and time changes. Einstein realized that time is no longer an absolute background stage on which events are played out, a domain unaffected by the events as Isaac Newton proposed. Einstein determined that time is derivable from physical processes and hence, affected by them. To explain it simply, time in Einstein's point of view became like a rubber band; time stretches and compresses depending if the objects studied are going relatively fast or slow in space. In my theoretical calculation of the paper titled "An investigation of Time in Relativity"—see the appendix, I took time and added to it an imaginary term. So now instead of having the letter T that describes time, we have $t=T+i\xi$. I took t and plugged it back into Einstein's equations to see if the Special Theory of Relativity would reject my added imaginary term of time. I found out that the imaginary term that I had added in the beginning to Einstein's equations did not have to vanish, which means that Einstein's equations allow the time to have an imaginary part. When I discovered this, I was in the library of Antioch University in Yellow Springs, Ohio. I felt very excited. At this time, I was somewhat isolated from the physics world of research. I was mostly teaching physics to undergraduates at Wittenberg University in Springfield, Ohio, which was about 15 minutes drive from Yellow Springs. Later on, I found out that Stephen Hawking had mentioned, in his literature, that physicists are using imaginary time in quantum gravity.

According to all of this, it seems that time can have an imaginary part. After discovering this in my paper, I sent it via email to my brilliant, Turkish friend, Bayram Tekin. At the time, he was doing Post doc research at Brandeis University in Massachusetts. I met Bayram Tekin when I was a graduate student at the University of Minnesota. I shared an office with him while he was working on his PhD in high-energy physics. He was one or two years ahead of me and about to defend his PhD thesis. We discussed physics all the time and if you made a mistake, he would insult you without thinking about the consequences. He cared less for how you feel. I remember once, I was working on a difficult QCD paper given to me by Arkady Vainshtein—Arkady is a professor at the Fine Theoretical Physics Institute of the University of Minnesota. Bayram looked at the paper and said, "You know, Sidi. I am gonna tell you something very interesting. This is a good paper, but for you to understand it, or if you want to really understand it, you need to rape it. You need to rape this paper."

Bayram Tekin was a devout Muslim, so when he told me to rape that paper, I was shocked. I knew that he was trying to tell me that I needed to completely understand the paper. The figure of speech was strange, that's all.

I sent my imaginary-time paper to Brandeis for Bayram to read. I was assuming he would respond in a day or two and insult me hard, but he didn't. And since he didn't respond right away, I figured he either forgot to read it or he found something interesting in it and was debating on how to respond. After two weeks or so, he responded and said "since you added an imaginary part to time, space has taken an imaginary part". The fact that I added an imaginary part to time caused space to react. Space and time, in the Standard Model, being related was a wake up call. Time and space are so tied together to the point that makes you believe that time is physical if you consider space to be physical, but again I found out in mathematics that pi is tied to numbers and always comes out in integrals. For example you would be working some mathematical problem where pi was not introduced at all and all of sudden pi pops out from nowhere. Pi is so tied to the natural numbers, but the natural numbers and pi itself are just creations of our imagination. They don't exist physically.

We can think of time and events as two parallel lines. One line is called the line of events and the other is called the line of time. We can assign a time "t" for every event. In fact, we can assign two, three or even more events to one time. Two things can happen at the same time; for example, two planes, A and B, can both land at three o'clock in the afternoon. But, can we assign two times to one particular event as measured by the same person? For instance, can Plane A crash at an airport at three o'clock in the afternoon on January 10, 2006 due to strong winds from a major storm, and also crash for the same reasons at four o'clock in the afternoon of the same day and have this be the same particular event? No one particular event, with specific conditions, can possibly happen at two different times. When somebody dies, they die at a specific time. John died at 3pm on Friday, June 3, 1956. In other words, we cannot assign two different times to the exact same event. Why can we assign one time to two events, but not one event to two times? This is a simple question, but it is hard to answer. Normally, we use time like a tape measure. We need to know when events occur in the same way we need to know the length of something. A tape measure measures a length, as time measures an event. We can assign two objects one length, but we cannot assign one object two lengths. For example, two rulers can have the same length but one ruler cannot have two lengths. There is parallelism between two ideas here. We created a tape measure to measure the length of an object. In the same way, we might have created time to measure an event?

Einstein found out that time stretches like a rubber band. If we stretch a tape measure, does the thing we are measuring change? No, the thing we measure doesn't change. What changes is our reading of the tape measure. When we use time to measure an event, does the event we are measuring change? No, what changes is the time we read on our clock. If we stick to this parallelism, we know that we created the tape measure to help us measure things. Is it true then that we created time to help us know when events occur? We know that the tape measure is not tied to the object we are trying to measure. If a tape measure were tied to an object, we would never know the exact length of the object, because if the object changed, the tape measure would change with it too. If we change time, we know that the order of events doesn't change. If the order of events changed, we would violate causality, which means you could be born before your mother, or the fallen dish could reassemble its shattered pieces and return to the tabletop. Of course, these violations do not make sense. The reason we cannot violate causality is because time and an event are two independent things, just as a tape measure and an object we try to measure are two independent things.

Sidi at Calpoly Pomona, CA

CHAPTER 7

Anything that Exist Physically in Space has Energy

We need to go back to our main question: Does time exist as a physical entity, such as matter, or is it a creation of our imagination such as money or the numbers 1, 2, and 3? We know that matter exists because it interacts with our bodies. A brick exists because if it hits something it can cause damage. I always tell the following to my students: You are not sitting on your chair; you are hovering over your chair. The idea of two particles touching each other is not as realistic as it seems. Particles interact through their fields. For example, if you take two magnets and point their north poles toward each other, you will notice that the two magnets repel each other or push each other away without touching one other. The two magnetic fields are interacting with each other at a distance. Now when you sit on a chair, the field generated by the particles of your body interacts with the field generated by the particles of the chair in, a somewhat of, the same fashion that the two magnetic fields interact. So touching is actually a vague word. I once asked Murray Gell-Mann about the classical vision that we have of particles. I asked if particles such as electrons protons are spheres that spin in space. He told me to forget the idea that particles are spheres, I said why? And he said the physics doesn't work. I didn't want to keep on asking, because he said, the physics doesn't work, with some discontent. The spherical model of the particle didn't work for him. He was probably on the side of the string model of the particle.

Energy can get extracted from matter and since matter exists, we can say that energy exists. For example, light is a form of energy that we know exists. There is this relationship between matter and energy. They are tied together. If one exists, the other has to exist too. Time on the other hand is not energy, or it does not contain energy. In fact, Planck constant, in the equation called the uncertainty principle, ties time and energy together. If time contains energy we can state that it physically exists such as matter does. But one can argue that things don't have to have energy to exist. Is there such a thing that doesn't have energy and exists physically? All the things that are objects and exist in this world have energy. Think of something that exists and does not have energy. A cup or a brick is made of matter, so they have energy. Light is made of energy. Numbers, ideas, and thoughts exist only in our head. They don't have energy. They don't exist physically, such as matter. We can use an inductive reasoning and state that all things that physically exist in space do have energy. For example, an electron exists in space; it has energy. A book exists in space; it has energy, and so on. One cannot find something that exists in space and doesn't have energy or made of energy. For one to negate this reasoning, he or she has to find a thing that exists in space and doesn't have energy. The problem is, there is no such thing. Finally one can state that time is not energy and it doesn't contain energy so, it must not exist in space since everything that exists in space has to have energy.

A physicist can say that the reason matter interacts with matter is because matter can change to energy and vice versa. So, an interaction can occur and an exchange of energy will follow. Things that don't have energy do not interact in the physical world. And if time is not energy; it will not interact in the physical world. We know that energy is stored in space. Is energy stored in time also? Is time stored in space? Any responsible physicist knows that time does not contain energy. Since it does not contain energy, it will not interact with a physical object that has energy.

Did we invent time or was time already in the universe before we came to existence? We know that we invented money because there was no money in the universe before we came about. We created money, out of necessity, to trade because we cannot possibly make every house or business item we need. One can make bread at home but cannot make a toaster.

Now, what about time? Did we create time out of necessity, like we did with money? Let's, for the sake of simplicity, assume that we created time. If we created it out of necessity, what was that necessity? Can we simply come up with a story to explain its existence? Imagine one neighbor wants to meet another neighbor to go hunt or fish. How can they meet? The first neighbor will tell the second neighbor, hey, let's meet to go hunt for bears, assuming bears were in that location. The second neighbor will not be able to ask when because time was not invented back then. The second neighbor will say "how" instead of "when". The first neighbor will tell him, when the sun is located there, in that region of the sky, we will meet here. The first neighbor will have to point to that region in the sky where the sun will come to. The second neighbor will say sure, I will be watching out for the sun, and as the sun gets to that location I will run here to meet you. The neighbors will keep on meeting at the specific spot, when the sun was at the specific location they both chose in space. Then one day, the first neighbor has something to do in the big part of day and wouldn't be able to come. He would go to the second neighbor and tell him, hey, can we meet "when" the sun is really close to the horizon?

The second neighbor will say, sure thing. And then one day, the second neighbor came up with an idea: How about we assign each location of the sun in space, a line, or a scratch on this here piece of goat skin. The first neighbor will say, good idea. And the clock was created.

CHAPTER 8

The Exclusion of Ether

I wrote this chapter with the help of two articles or two books which I lost. I could not locate them when I was putting this book together. This book was written in the span of eleven year and I moved several times. In the past, scientists thought that light needed a substance called ether so that it can propagate in space. They figured that since sound needs air to propagate, light might as well need ether to propagate. When you say the word, go, for example, your vocal cords vibrate and exert an amount of pressure on air. This pressure travels in air, in somewhat, the same fashion a wave travels in a guitar string. The wave propagates from the point where you plucked the guitar string to each end of the guitar string. The wave in the guitar needs the medium, the string, the same way the wave of the sound of the word "go" needs the medium, air, to travel to the ear. If there were no air, there would be no sound for the word "go", and if there were no string in the guitar, there would be no sound of the guitar music. The reason you hear music from the guitar is because when you strike a guitar string, the string vibrates and beats on air and creates pressure in it, the same way your vocal cord creates a pressure in air when you speak. But, sound needs air to travel. Knowing this, scientists speculated that light does the same thing: it creates some sort of pressure in ether and this pressure propagates in space at the speed of light. According to this picture, electromagnetic fields were seen as deformation in this medium. The trouble was that the ether seemed to have a boundless thermal capacity. Which means, it can absorb heat with no such limit. And since nothing could prevent normal matter from increasingly yielding all its heat to the ether, in the form of electromagnetic vibrations of high frequency, this will give rise to an instability, and this instability means that material bodies should be unable to retain heat, or to remain in thermal equilibrium with their environment, in obvious contradiction with experimental evidence.

It was suggested that this substance called "ether" was present everywhere, including the "empty space", and light waves traveled through this ether at the speed of light. However, different observers, moving relative to the ether, would see light coming at them at different speeds since ether would be carrying the wave. The way to validate this, is to measure the speed of light as the earth moves through the ether on its orbit around the sun. So, light speed measured in the direction of the orbit of earth (when we are moving towards the source of light) should be higher than the speed of light measured at right angles from the direction of the motion of earth (when we are not moving towards the source of light). This experiment, called the Michelson-Morley experiment, was performed in 1887. To their great surprise, it was shown that the speed of light was exactly the same in both cases, which disapproved the effect of ether. If it existed like air does for sound, it would have an effect on the speed of light the same way air has an effect on the speed of sound. The speed of sound depends on the temperature, on the density of air and also on its resiliency or elasticity, called the bulk modulus. Disturbances are transmitted through a gas as a result of collisions between the randomly moving molecules in the gas. The transmission of a small disturbance through a gas is an isentropic process. The conditions in the gas are the same before and after the disturbance passes through. Because the speed of transmission depends on molecular collisions, the speed of sound depends on the state of the gas. The speed of sound is a constant within a given gas and the value of the constant depends on the type of gas (air, pure oxygen, carbon dioxide, etc.). Water is more elastic than air; this is why the speed of sound is larger in water than in air. Steel is more elastic than water; this is why the speed of sound in steel is larger than speed of sound in water.

The Native Americans knew about this. For example, when they expect a horse rider coming towards their camp they feel the ground for sound instead of listening to it in air. Now, why did I bring the physics of sound and its medium to this discussion? The reason is one can see the effect of the existence of air on the speed of sound. Air is a medium that exists that has an effect on the speed of sound. If ether had exited in empty space it would have had an effect on the speed of light, but it turned out that it doesn't exist in empty space so it doesn't have an effect on the speed of light. The point of all of this is we need to know if time has an effect on a physical process. If it has a physical effect that might mean that it exists physically. The best possible solution is to know of an experiment that will prove or disprove the existence of time, like the Michelson-Morley experiment did with ether. To come up with an experiment that will disapprove the existence of time, one first needs to know the effect of time on physical things. I have an experiment in mind but I cannot state it in this book for fear that some experimentalist might do it before I do it myself. This has happened to me before. I wrote a paper in High Energy physics called the Dissociation of Upsilon in a Quark Gluon Plasma. Two scientists—I don't want to mention the names—used a formula from my paper without giving me the credits for it. If it weren't for Joseph I. Kapusta who had caught it, the two scientists wouldn't have mentioned my name in their paper. Thankfully, they mentioned my name at the end.

CHAPTER 9

Can the Brain Understand Itself?

Is it possible that we came up with the idea of time the same way we came up with the idea of ether? Did we introduce time to help us understand the physical world around us, the same way we introduced ether to justify the propagation of light in space? This could be possible because so far we don't have any experiment that proves otherwise. It is truly hard to know things the way they are, or to see them the way they really are. In my opinion, it feels like our eyesight is blurry and even though we are wearing glasses, we cannot see things as clearly as the way Nature "sees" them. As we live and learn, from generation to generation, we are putting on better glasses and one day we will see things as clear as the way Nature "sees" them. Even though we are a part of Nature, we do not see things as clearly as Nature "sees" them or "feels" them. We see things with our eyes and we feel them with our senses. Our eyes can only see things that we need to see to survive, such as an apple, water, or a piece of bread. We cannot see things that we don't need to see to survive, such as quarks or air molecules. Things we can see are not the only things in the world. It is possible that there are things that exist in this world that we don't know about them. Our ancestors didn't know about Black Holes when, in fact, they did exist. Our sight is limited because our eyes are just organs. Our hearing is also limited. Are our thoughts limited because our brain is an organ too? Yes, it is true indeed. It is very hard to judge a thought. We don't have an absolute frame of reference to measure a thought; the same way we don't have an absolute frame of reference to measure the exact speed of an object. When you are traveling at 75 mph you are just traveling at that speed with respect to the ground. The ground is not an absolute frame because the earth moves with respect to the sun and the sun itself is not an absolute frame because it moves with respect to the center of our galaxy, the Milky Way, and the galaxy itself move with respect to the center of the universe and we don't know where the center of the universe is. When we judge a thought we just compare it to other thoughts. We don't have a perfect scale to measure a thought, but we do have science.

We can use science as a scale to weigh a thought, but even science itself is based on some thoughts that need to be weighed in order to find their exact strength or truthfulness. It is possible that our organs are made to the limit of our survival. Nature will not give us more than what we need, something I learned in physics, in regards to the conservation laws of physics. For example, energy is one of the quantities that are conserved. Nature conserves things and it probably gives us only the things that we need to survive as species. Our thoughts are probably constrained and free only to the limit to help us survive. Our mind probably cannot understand things we don't need for our survival as species. Richard Feynman taught of this in some way. He left a statement written on his chalkboard when he died. It said: what we cannot create we do not understand. Some people might argue this by saying, we didn't create ourselves but we understand ourselves. And other people might say that we don't really understand who we are and therefore, will agree with Feynman's statement. Do we really understand ourselves? Do we understand our brain? Do we understand how our brain generates thoughts? Can the brain understand the brain?

Understanding, in itself, needs to be questioned. Werner Heisenberg once said "the exact sciences start from the assumption that in the end it will always be possible to understand nature, in every new field of phenomena, but that we may make no a priori assumptions about the meaning of the word understand".

I once was at a conference in the physics department of the University of Minnesota, my Alma Mater, listening to a scientist talk about the brain and neurons and their interactions. We were eating cookies and drinking coffee in the lounge. I approached the brain scientist and asked him, "Can the brain understand itself? We are searching the brain with our brain. Can the brain really know itself?" The brain scientist looked at me and started to think. After pondering awhile, he finally said, I don't know. It is a very good question, though. The I-don't-know answer is good and bad at the same time. It is a good answer because it doesn't mislead you and it is a bad answer because it doesn't tell you anything. There is no information in the I-don't-know answer, except that the person who said it doesn't know the answer. If we cannot see things as clearly, can we really understand them? What we need to look into is how we understand things. We didn't create our world, so can we understand it? We probably understand only things that are indispensable for our survival. Now to discuss what "understand" means can be a book in itself. I truly recommend somebody to write a book to explain the definition of the word "understand". Galileo responded by saying, "God has written; now we must read" (page 35 Isaac Newton James Gleick). God gave us the book, which is the universe, and we need to read it, but the book cannot be understood unless one first learns to understand the language and read the letters in which it is composed. It is written in the language of mathematics. You can see here that every body is telling you something: One is telling you to read; another is telling you it is hard to understand, because we don't know what "understand" is, and another is telling you to forget about understanding things; just get used to them as you go. Richard Feynman once said this latter to his students.

CHAPTER 10

My Meeting with Nobel Laureate James Rainwater

P lato, like I stated before, was the first to imagine the idea of timeless existence and to emphasize it as a reality. One might ask: "Can you give me an example of something that is timeless?" One thing that comes to mind is a mathematical truth. It does not come into being when we discover it, yet its discovery is a very real event. You can argue that real numbers, for example, have no physical existence. They are a human creation and thus a product of biological and cultural creation. Let's find an example of a mathematical truth that is somewhat simple and self-explanatory. By adding consecutive odd numbers, you always get a perfect square number (e.g. 1+3=4; 1+3+5=9) 4 is the square of 2 and 9 is the square of 3, which represent the amount of odd numbers we added in the two examples. This mathematical truth is always true. It doesn't need time to exist. It doesn't need anything to exist. Of course, when we think about it, we realize that it exists, but it was there in the void before we came to existence.

This brings me to a story. When I was an undergraduate student at Columbia University in New York City, one of my professors was the Nobel laureate James Rainwater—He received the Nobel Prize for the discovery of the connection between collective motion and particle motion in atomic nuclei and the development of the theory of the structure of the atomic nucleus based on this connection, which in some skewed way, means he showed that the nucleus is not really a sphere, but somewhat a flat sphere, like an egg, for an exaggerated example. He was a weird professor in the good sense. I had to take him for some laboratory courses to complete the requirements for my undergraduate degree. James Rainwater was an old-looking man who wasn't really relatively that old at that time. He was probably 66 years old, but looked like he was over 80, except for his eyes; they were sharp and full of life. His poor saggy, baggy body reflected wear and tear, probably because he spent most of his adult life in the stagnant air of his laboratory, embedded in a magnetic field, or an electric, bombarded by particles and radiation. Enrico Fermi, the famous Italian-American physicist had paid dearly for being in lab all the time, bombarded by particles and radiation

Like James Rainwater. He died of stomach cancer, and to add insult to injury, two of his graduate students who assisted him in working with radiations died of cancer too. Fermi and his students knew that such work carried a huge risk of getting sick, if not dying, but they cared less for their safety and they forged ahead with their research, for they considered the outcome to be so vital to society.

Anyway, I went to James Rainwater's office on the second floor, if I remember the floor correctly, of the physics building to introduce myself and ask him questions about the class I would be taking. If you saw him, you would think he sold gasoline in a gas station—he had shabby clothes on and he has a belt that was secured up high tightly around his belly making his belly protrude in a strange way. He walked very slowly and used a wooden cane. I entered his office and saw that it was full of electronic gadgets, old dusty instruments, and old faded pictures of long-gone students. The shelves were full of physics books, some of them were shelved vertically, some horizontally and almost all of them were out of print. The papers were yellow and so full of dust that if you opened one of the books, you would probably sneeze. There were piles of yellowish papers stacked all over the place.

I said, "Professor! I am going to be your student this semester. Do you mind if I ask you some questions?"

"Before you ask me any question, I have to take a picture of you," he responded.

He took pictures of his students because he couldn't remember their faces when he would grade their lab reports. He mostly judged them, not just by how they wrote their lab reports, but also by how they handled lab equipment and how they related their results to the meaning and purpose of the experiment and so forth. He took a picture of me and said, "Now you can ask me questions."

"How do you prefer the lab reports to be written," I asked.

"You choose the way you want to write your lab report, and I'll correct you when I see something wrong. Never ask me any question before you think about the answer! I don't like people who don't use the operator, Think."

He stood up and wrote the word, THINK, with a Chinese hat on top of it—-an upside down V. Physicists indicate operators in such a fashion. He said, "You should always use the operator, THINK."

Professor Rainwater had a heart problem. My guess was it developed from regularly exposing himself to radiation. He had lived through many strokes in the past. He was tenure with the university, he had the Nobel Prize, and he got all the prestige most people dream of. He could just stay home and play chess with his grand children if he wanted to, or go on vacation to Hawaii, or volunteer at a home shelter serving soup, but instead he kept on coming to the physics building, which was really an ugly building in the inside, still is. It usually took him twenty minutes to cross the campus and ten minutes to go up to the second floor when the elevator was not working. One day I saw him on the campus floor, lying on his back while a young man was working on him, pushing his chest, and trying to resuscitate him. The ambulance came shortly afterwards and took him away. I tried to get hold of him after that day, before I had to leave to the largest university in the US, at that time, the University of Minnesota. One day I called the secretary at Columbia University physics department to see if I could get a recommendation from Professor Rainwater, but the secretary, Sheila Page, if I am correct, told me that he had passed away. This story might seem irrelevant to the subject of this book, but I felt the need to write it to inform you of the operator, THINK, and the following operators.

There is another operator called C, or charge conjugation operator, that we need to understand. It is not a good symmetry in Nature. It is something of a misnomer, for C can be applied to a neutral particle, such as the neutron (yielding an antineutron). It also changes the sign of all the "internal" quantum number—charges: baryon number, lepton number, strangeness, charm, beauty, truth—while leaving mass, energy, momentum, and spin untouched. Charge conjugation is not symmetry of the weak interaction either. When applied to a neutrino (left-handed), C gives a left-handed antineutrino, which does not exist in Nature. If it acts on a particle, then that particle becomes the anti-particle of the original operator. C doesn't change the mass; it only changes the charge. So if it acts on an electron then that electron becomes a positron, which is an anti-particle to the electron. If the positron collides with the electron then both the electron and positron will disappear and all you get is radiation without any mass whatsoever. There is another operator called P, or parity operator. This operator acts on a particle like a mirror "acts" on an object. If something turns clockwise in the real world, it will look like it's turning counterclockwise in the imaginary world, which is behind the mirror. If it acts on a left-handed neutrino, it changes it to a right-handed neutrino. This parity operator is violated because the right-handed neutrino has never been observed. What's so interesting about this P operator is that when it joins C, it forms a composite operator called CP. If C acts on a right-handed antineutrino that exists in Nature, it changes it to a right handed neutrino that doesn't exist in Nature. But when P acts on this right-handed neutrino that doesn't exist in nature, it will change it to a left-handed neutrino that does exist in Nature. So this makes the CP operator symmetric. Please bear with me talking about these operators, THINK, T, P, and C, because they will come in handy in some skewed way later on as we dig deeper into this book.

CHAPTER 11

Taking Time Away From Speed

There are many physicists who want to create a new physics, which is devoid of time. Now, how can we describe speed without using time? You can tell if somebody is moving faster than somebody else without using time. The person who is faster arrives at the destination first. If you look at them you will notice that the legs and arms of the faster person are moving and pumping faster. If we can tell that one person is moving faster than the other without using time that means that speed can be described in another form, or a formula that will not include time. Now somebody might ask us the question: how fast is the faster person moving? This can be a difficult question to answer if we don't include time. We can find a unit of speed and write each speed in terms of this unit. Physicists have already done this. For example, the mach is a unit that describes the speed of sound. One mach is about 330 meters per second, depending on the temperature. The speed of sound in water does not depend on time. It is a physical process, or an interaction like Boltzmann had stated. In physics there is a formula that ties speed to time. There is also another formula that ties speed to acceleration without the use of time, but one would need to know the acceleration, and the acceleration itself depends on time and it can be described as the change of speed with respect to the change in time. Another way of avoiding the use of time is by picking a certain speed that we know about in nature and use it as a reference or a factor, like I did with the mach, in order to write down any speed we might find in terms of this chosen known constant speed. Any speed in the universe can be written in terms of this speed. For example, we can use the speed of Earth with respect to the Sun and call that speed, Earth Speed with Respect to the Sun and we give it the acronym, ESRS. We can also use the speed of the electron of a stable hydrogen atom and use this speed as a reference like I did with ESRS and mach. Now all the speeds we know can be written in terms of this speed. The speed of light can be written in terms of either the speed of Earth or the speed of an electron. So far, I didn't do anything useful in finding the speed without the use of time. All I did is state that any speed can be written in terms of the speed of Earth around the sun, or in terms of the speed of sound. Now, all we have to do is find a way to dissociate Earth speed from time. If we find that Earth speed is dissociated from time or doesn't need time at all, this will lead us to believe that all other objects' speeds can also be devoid of time. We can state that when Earth goes around the sun it cuts into space and space is the only dimension involved in this process. Of course, somebody might ask the question, how fast does Earth cut into space?

I remember when I was a graduate student attending the University of Minnesota, I shared an office with my friend, Ramin Daghigh, who was also a graduate student at the time. He was working on his thesis in Astrophysics and I was working on mine in high-energy physics. We both worked extremely hard. We were always at our desks talking physics and solving problems. We had one large office, but it was shared with about four other graduate students. One day I told him that everything we do is a waste of time. He asked how come? And then I repeated to him that everything we do in life is a waste of time. He criticized my statement. A year or so later, he sent me an email. He was now working at the University of Arkansas. In his email, he stated that I was a genius. He said I was a genius because I had once told him that everything we do is a waste of time. Now is it true that everything we do is a waste of time? From one point of view, every time you save some time, you would lose it anyway. If you do something or you don't, you will lose that time. You are always losing time no matter what. At the end of the day you would lose your twenty-four hours regardless of what you did on that day. You could have made a business transaction and made two hundred thousands dollars or you just slept through the whole day. But, regardless of how much you make or how much you sleep, at the end of the day you will lose your twenty-four hours. You might say that you didn't waste your time because you made some money, but you need to realize that in a smaller scale you made some money, for sure, but in a larger scale, Nature sees that you didn't do anything because money is just a human consideration. To Nature, money doesn't exist; it exists only in our mind. To Nature, whether you made money or not you are still the same. From the other point of view, anything we do is a waste of time because at the end of our lifetime we will cease to exist regardless of what you did in your life. You might say this is dangerous because if people believe this, they would just sit and do nothing. If you do something with your life or do nothing with your life, in the end you will lose your life anyway. If, for example, you accomplished some good deeds, and another person didn't do anything good, at the end of your life you will both die and be the same. When you die, you will not feel anything and neither will the other person who didn't do anything with his life. You spent your time giving back to the community and the other person spent his time sleeping and in the end you will both lose your time. If you write a book, that book will one day be out of print—like this one. If you build a building, one day that building will get old and it will have to be demolished to build a new one. If you own a piece of land, one day you will lose it because one of your grandsons will sell it. The land will always be there, but many people will own it before they die, but the land doesn't die like humans do. My point is whatever you do in life will end some day. One might say that human procreation will not die. I would say that dinosaurs procreated and ceased to exist. Humans one day will cease to exist as species. I once asked myself why do we have to die? Can we live forever? Scientifically this might be possible, I said to myself. All Nature needs is energy and intelligence and it will do anything for you. As long as you don't violate the laws of energy, which means you won't try to get energy from the void, you will be fine with Nature.

When I was teaching at Wittenberg University in Springfield, Ohio, I met David Mason, a biology professor—a great guy. We used to meet at the gym and talk about science. I asked him, could we avoid death? He said we have to die so that the human species can live. I said, what? He explained, nature is changing and we have to change with it. If we live longer, our body will not change with nature and when we produce an offspring that offspring will not have the changes stored in it and nature will kill him or her because he or she would not be compatible with nature at that specific time; he or she will not have the change that occurred in nature stored in his/her genetic material. That's interesting. We die so that others can live. This could be the solution for our survival as species, but again who knows? One day something will happen and our species will get exterminated. At the end of the "day" our species will die. So whatever we do is a waste of time because in the end we will die. Our children will die and their children and this will cascade down to the last generation of offspring. Another explanation for this is that whatever we do is a waste of time because it is probably true that time does not really exist physically. If something doesn't exist, wasting it or not wasting it is the same.

CHAPTER 12

Physicist vs. Philosopher

According to a typical dictionary, reality is the fact, state, or quality of being real or genuine, that which is real; an actual thing, situation, or event. For example, when a person dies he or she has experienced the reality of death. Death is a reality as life is a reality. When a child is born, it is real. Humans define reality as what is written in the dictionary. It is possible that this reality is not the true reality, or the absolute reality. Reality is a perception, and we only perceive what we experience in life, but there are many things to experience, and we cannot experience them all because we die. When we look at the stars in the sky, we tend to think they live forever. When stars look down at us, they see flashes of "light" that live for a millisecond and then die. To stars, we live only a very short time. Life allows us to leave a little bit of what we experience or what we know for the coming generations. We don't live long enough to watch the whole game of life and fully understand its rules. We only know a few rules, which help us understand only a portion of the game. An analogy of this would be if you were to watch a movie, but never finish it. You would tell your children of what you know about the movie and your children tell their children about that same portion. So, the coming generation can never know the ending. They can only speculate. This is what is happening to us. We don't live long enough to know the mystery of life.

You know when you ask, "What is reality", it is like asking "What is the truth". If you are religious you might easily be convinced about the truth, but if you are not religious and you want to know the truth you will encounter a difficult task. When you ask the question, "what is reality", it comes out from your mouth. Your brain automates your mouth, and your brain is programmed by Nature. Every cell in your body is put together according to natural laws. The matter that your brain is made of is from Nature. The laws that put the molecules together to make your brain, and make you pose these questions, are natural laws, or physical laws. It is like Nature is speaking through us, we have no personality of our own. When you ask, "What is reality", you are automatically giving an existence to it.

When scientists want to know if something is real, they calculate the probability for it to exist. If they find the probability to be zero, they conclude that the thing doesn't exist or it is not real. Physicists are somewhat different from philosophers. Philosophers are allowed to give existence to things even if that thing doesn't exist physically or is not real. Scientists are not allowed to do that. A scientist can sometimes get away with things, but only for a while before an experiment comes in and claims that that thing doesn't exist physically or is not real at all. A little story goes like this: A philosopher and a physicist were talking about the existence of things. The philosopher said, "For example, this stick here on the ground doesn't exist physically. It only exists in our mind."

"The stick exists, my friend. It is real." The physicist said.

"I am telling you," the philosopher responded. "It's possible that it doesn't exist."

"I can prove to you that the stick exists," said the physicist.

"I know what you are going to do," said the philosopher.

"Of course you know."

The physicist picked up the stick and hit the philosopher on his head.

"Even if you hit me with it," said the philosopher scratching his head. "It doesn't mean it exists."

The physicist thought for a while and rest his case. There was no way he could convince the philosopher. The physicist did an experiment on the philosopher by hitting him on the head and from that experiment he concluded that the stick had interacted with the philosopher's head. Philosophers are always allowed to wander and think about non-existent things, which is good sometimes because that allows freedom of thinking.

Going back to the cubes and the container I discussed before in chapter 5, if for example, you don't put any green cubes in the container, the probability of finding a green cube in that container will be zero. If you put one red cube in a container that has 9 white cubes, the probability of pulling out a red cube from the container is 1/10. If you put 19 white cubes, the probability changes to 1/20, and the more white cubes you put in the container, the less probability of pulling out the red cube from the container. What will happen if we put a billion white cubes in the container? The probability will be extremely small, almost one billionth, but there will still be a chance of pulling out the red cube. For us not to find the red cube, the probability has to be zero. This is exactly what scientists do when they want to know if something exists. They calculate the probability and from that probability they determine if the thing or a particle exists. And to do that, they need to know about the particle and its behavior, which means they need to know the wave function of the particle and then from that wave function, they calculate the probability density. Now you ask yourself what the heck is this probability density? Let me explain it in a simple way. If we understand the number density, we can understand the probability density. The number density is the number of things you are interested in counting, divided by the volume where these things are. If you assume you have 10 books in a room, which is 10 meters high, 10 meters long, and 10 meters wide, the number density of books is 10/(10X10x10), which is 1/100 books per meter cubed. So it is just the number of books divided by the volume. We can use the same principle for the probability density, the only thing we need to do is, replace books by the probability. I explain this to my students all the time. I pick one student from the classroom and I tell the other students that we are going to find where this student is located without seeing him. We will do this using the probability density. Now I say, imagine we have some cloud of dust in this room. The dust particles represent probability particles. When you look inside the room you see this distribution of dust particles or probability particles floating in the air with the majority hovering or hanging in the location where the student is sitting. So in this case, you can tell where the student is sitting without watching him or her sitting.

Now this probability density is somewhat confusing when you start talking about particles and wave function. In quantum mechanics a particle is governed by a wave function. It is a mathematical function like for example $f(x)=x+2$. What does this function do? If you give it a number, say 4, it adds to it 2 and results in 6. If you give it the number 10, it adds to it 2 and results in 12. This is the function of this function. You need to know that the wave function for a particle is more complicated than the one I gave you. Knowing the function you can have some information about the particle. When physicists want to calculate the probability of finding a certain particle, say an electron, in a small volume, they get the wave function and multiply it by its complex conjugate and then sum over the whole volume of interest. What's so interesting is that probability is a conserved quantity. When physicists try to calculate the probability they have to normalize the wave function, which means they have to state in a mathematical form that the probability of finding the particle in the universe is 1, which means that the particle exists in this world. For example, the probability of finding your cousin in the world, assuming she's alive, is 1. Physicists assume that no particle can escape our world, so they conclude that probability is like energy; it is conserved. Now, nobody can explain why energy, parity c, and momentum are conserved quantities. All we do is assume that they conserve until somebody does an experiment and tells us otherwise. This is exactly what happened with parity. Physicists used to think that it conserves until, Lee, yang, and Chu, I talked about before, proved that it doesn't. Now how is this related to the subject of this book, which is time? Well parity is an operator and Time can also be an operator, so if we proved that parity is violated, that means that it probably exists in the physical sense. Thus, if we prove that Time reversal operator is violated, I think that will probably show us that time does physically exist.

CHAPTER 13

Time Reversal

Reading the physics literature, I found out that an increasing number of experiments at CERN (Conseil Européen pour la Recherche Nucléaire), SLAC (Stanford Linear Accelerator Center), and Brookhaven labs are confirming the violation of time reversal invariance T. If T is really violated it will signify a fundamental asymmetry between the past and future. The processes, which violate T symmetry, can cause destructive interference between different paths that the universe can take throughout time. This interference can eliminate all paths except for two that represent continuously forwards and continuously backwards time evolution. A long-standing problem of modeling the dynamics of T violation has been in existence and physicists could not, so far, pin down if its effect is physical. I once was told by the Russian physicist Michael Voloshin, who now teaches and does research at the Fine Theoretical Physics Institute of University of Minnesota, that if T is violated it will give existence to particles and antiparticles that have different mass. So far, particles and antiparticles have the same mass. For example, the electron and its antiparticle, the positron, have the same mass. Another example, the up quark and the anti-up quark both have the same mass. If T violation can generate a physical effect, this will allow us to state that time is probably physical. Physicists have tried to find a physical effect but were unsuccessful. They even tried the discrete time. Since energy is discrete, and energy is loosely tied to time in the Heisenberg uncertainty principle, they assume time can also be discrete. I remember when I was an undergraduate student at Columbia University, Norman Christ, T D Lee, and Richard Friedberg were all working on discrete time. I looked in the physics literature these days; I didn't find a thing that could convince me that discrete time is physical or has an effect on physical things.

Richard Friedberg once taught me classical mechanics at Columbia University. He always came to class with a sweater full of hair flakes and lint. He looked like he slept in his clothes. He also had these brown short boots that he wore all the time. One of the boots had a broken zipper that Richard fixed using a safety pin. I always asked myself, what happened to the salary he gets from Columbia. The clincher, one day he came to class with a cookie in his hand and when students entered the class and were taking their seats, he went behind the open classroom door and stood there eating his cookie, hiding from us. We could see his short boots under the door.

Richard Friedberg was one of the brilliant physicists at Columbia, if not the world, but he was weird. I liked him a lot because he was absent minded and cared less about what was going on around him. I once went to his office to discuss this problem of getting the energy from an electron without affecting the motion of the electron. I told him that if we kick an electron with an initial velocity, the electron will move with some kinetic energy and by moving it will generate some magnetic field. If we take some energy from that magnetic field, will the electron hit its target with the same kinetic energy we gave at the start? He said no. I asked "why". He said when you take the energy from the magnetic field you will affect the speed of the electron. This interaction of one trying to get the energy from the magnetic field, which affects the motion of the electron, shows the physical existence of the magnetic field. Can time interact with an object in such a fashion that it shows its physical existence?

When I gave an imaginary part to time in my paper, I wasn't aware of what would happen to space. At that time, I cared less about space. Einstein stated that at high speed, time gets dilated and the distance gets contracted. Let's say, for instance, that you take a bunch of people with a maximum lifespan of 130 years, and you send them through space to a planet that is 135 years away—this is the same as when somebody asks you how far is the gas station? You say 10 minutes away. If you ignore relativity, all the travelers will be dead when they get to that planet; but if you don't ignore relativity, which means you include it in your calculation, there would be some traveler who would get to that planet. The reasoning behind this is that they went very fast. So, the distance that they traveled, as seen from their reference point, got contracted due to the relativistic effect. To us who are on Earth, as seen from our reference point, the time that they had lived got dilated so they lived longer and that allowed some of them to get to their destination alive. So, you can see here that the relativistic effect is shown in both space and time. So which one is right, the relativistic effect on space or the relativistic effect on time?

Now this same phenomenon has occurred to the muon particles that get formed in space and travel down towards the Earth at high speed. These particles are produced at the outer edge of the Earth atmosphere by incoming cosmic rays hitting the first traces of air. They are unstable particles, with a "half-life" of 2 microseconds (2 millionths of a second), which means that if at a given time you have 100 of them, 2 microseconds later you will have 50 left, 2 microseconds after that 25, and so on. So if it gets created high up in space, it will decay when it gets to the surface. It actually decays to an electron and neutrinos. By doing the experiment, one finds out that some of the muons makes it to the surface of the Earth. The muons survived because of the fact that there was a relativistic effect occurring during the travel. Much less time passed in their frame of reference. To us on Earth, the muons have lived longer, but to them in their frame, the distance they traveled has shrunk. Could we say that the reason they reached earth is because the distance they traveled got contracted due to the high velocity they were approaching the Earth surface with, and ignore the time dilation? Could we stick to the effect on space and ignore the effect on time?

CHAPTER 14

Aging and Time

If an observer, far away from a large star that is collapsing to a black hole, is measuring the time it takes for the star to collapse to a black hole, he or she will find that their measured time is dilated and as the particles of the stars are speeding towards a point at the center of the black hole or a singularity, the time will get even more dilated, to the point that it will take an infinite time for the process of collapse to be completed. But according to the observer connected with the collapsing star, this would be completed in a finite time. The time it would take for the star to shrink to its Schwarzschild radius would be infinitely extended according to an external observer. Schwarzschild radius is the distance from the center of the collapsing star to a line that lies below, where not even light can escape. The light inside the Schwarzschild radius will try to get out but the space inside is stretching, making it hard for the light to come out. It will be like a person running on a road while the road is stretching underneath; such as the finish line is moving away faster than the runner can. The runner will never reach the finish line. This is what happens to the light inside the collapsing star.

Now since it takes an infinite time for an external observer to see a star collapsing to a black hole, how come they exist in space? Since the space inside the black hole is stretching, not allowing the light to escape, it could mean that time is irrelevant, which means that what's happening to the collapsing star has nothing to do with time. It has to do with the physical processes that are occurring inside, such as the gravitational forces and the pressures derived from the heat of the squeeze. Like I wrote in one of the previous chapters, our life span has nothing to do with time, but instead, it has to do with the physical processes occurring in the cells of the body. We age because of the physical and biological processes occurring inside the cells, not because of time.

As Karen Wright stated in her March 5, 2012, Special Collector's Edition of the Scientific American article, Times of Our Lives, scientists in search of the limits to human life span have traditionally approached the subject from the cellular level rather than considering whole organisms. So far, the closest thing they have to a terminal timepiece is the so-called mitotic clock. The clock keeps track of cell division, the process by which a single cell splits into two. The mitotic clock is like an hourglass in which each grain of sand represents one episode of cell division. Just as there are a finite number of grains in an hourglass, there seems to be a ceiling on how many times normal cells of the human body can divide. In culture they will undergo 60 to 100 mitotic divisions, then call it quits. "All of a sudden they just stop growing," says John Sedivy of Brown University. "They respire, they metabolize, they move, but they will never divide again."

Cultural cells usually reach this state of senescence in a few months. Fortunately, most cells in the body divide much, much more slowly than cultured cells. Eventually, perhaps after 70 years or so, they too can get put out to pasture. "What the cells are counting is not chronological time," Sedivy says. "It's the number of cell division."

In the physics of the black hole, I can safely assert that the process of condensation inside is independent from the chronological time. In our sun, for example, there are two major forces which are working against each other: The force of gravity which is due to the particles of the gas inside, pulling each other in; and the forces that come out from the "cooking pressure" of the sun. The hydrogen gas is interacting inside, or cooking inside, so to speak, and causes the medium to heat. This increase in heat causes a pressure that tries to send the particles out away from the sun and into space, but since gravity is strong it traps these particles; but if the star is huge the gravitational force will overcome the pressure force and, at later time, when most of the hydrogen gas is fused to helium, the star begins to collapse on itself. This process is independent of the chronological time. The chronological time is just a tape measure that measures the times between events occurring inside the black hole, the same way you use a tape measure to measure your kitchen table.

G.T. Whitrow stated in his book, the Nature of Time, page 124, that Minkowski's concept of time has proved to be one of the most valuable contributions ever made to theoretical physics by a mathematician. In his enthusiasm, Minkowski exclaimed "Henceforth space by itself, and time by itself, are doomed to fade away into mere shadows, and only a kind of union of the two will preserve an independent reality." This famous, but excessive, claim tended to reduce the importance of time much more than that of space. Herman Weyl seemed to agree with Minkowski. He expressed that the passage of time is merely to be regarded as a feature of consciousness that has no objective counterpart. In his case, the four-dimensional continuum is neither time nor space, the guiding concept is evidently more spatial than temporal. Einstein was more aggressive compared to Minkowski or Weyl, for he came to the conclusion that objective world of physics is essentially a four-dimensional structure.

Parmenides submitted the ideas of becoming and perishing to acute criticism and concluded that time does not pertain to anything that is truly 'real', but only to the logically unsatisfactory world of appearance revealed to us by the senses.

CHAPTER 15

One of Einstein's Biggest Mistakes

There are things that don't physically exist, but they can play the role of something that exists physically. This can affect our thoughts and make us believe that they really do exist physically, when in fact they don't. This concept often occurs in science. For example, in the famous Halliday and Resnick physics textbook, suppose that from orbit we watch a race in which two boats begin on the equator of the Earth with a separation of say, 40 km and head due south. To the sailors, the boats travel along flat, parallel paths. However, while navigating the boats, the sailors will get closer to each other until eventually their boats touch. Here, we assume the South Pole is covered with water not ice. The sailors can deduce that the boats reach each other because some sort of force has been acting on them and the boats. We, on the other hand, can see that the boats touch each other because of the curvature of the Earth's surface. We can see this because we are viewing the race from "outside" that surface, looking over. So, the force the two sailors talked about doesn't exist physically, it exists only in their mind as a concept; and what's so dangerous about this is that it makes sense to them, but it takes somebody to be "outside" the surface to see that it really doesn't make sense. These two sailors might develop a theory using this non-existing force and it might work for them for a while or for a long time, but eventually, when they want to improve their theory or their force and try to combine it with other forces, they might encounter some problems.

How can we avoid this kind of situation? Einstein himself was drawn into somewhat the same situation. He was led to believe that there was a force that holds the stars in a fixed position, the same way the two sailors believed in the attractive force that attracted their boats towards each other. Einstein was influenced by the idea that the stars in the sky are static. So, when he developed his general theory of relativity, he thought he encountered the same problem Newton did: his equations showed that the universe is either expanding or collapsing, yet he stuck to the idea that the universe was static and inserted a constant term to the original equation, called the cosmological constant. This constant cancelled the effects of gravity in the physics of large scales and led to a static universe. After Edwin

Hubble showed Einstein the red shift, the Doppler effect on the lights coming from the stars, he was convinced that the universe was expanding. Later on, Einstein called the cosmological constant his greatest mistake.

The important thing is to understand what is going on so not to fall into this kind of situation. Marie Curie once said, nothing is to be afraid of; it is just to be understood. But understanding in itself is a big thing. It doesn't come easy. I was once tutoring math to a little girl and she started to cry because she could not understand what I was explaining to her. It was easy for me because I understood, but it was extremely hard for her because she didn't understand. Understanding doesn't come easy; it takes a lot of work. Sometimes people do things without understanding why they do them. It is possible that we are making a big mistake in our scientific research and we don't know about it. Animals also do things without understanding them. In the famous Halliday and Resnick physics textbook, I read that the giant hornet preys on bees. However, if one of the hornets tries to invade a beehive, several hundred of the bees quickly form a ball of bees around the hornet to stop it. The hornet is inside the ball of bees trapped. After about 20 minutes the hornet is dead, although the bees do not sting, bite, crush, or suffocate it. Why, then, does the hornet die? The bees don't understand how this mechanism works but they do it anyway. The key concept here is that, because the surface temperature of the bee ball increases after the ball forms, the rate at which energy is radiated by the ball also increases. All this energy or heat has to be absorbed by the hornet and the hornet cannot withstand all this rate of energy, thus it dies. Now, the bees lose an additional amount of energy to thermal radiation. We can relate the surface temperature to the rate of radiation (energy per unit time).

So, it is possible that time has been created in our imagination to help us understand the physical world around us—the same way we created money to help us trade. It could be an imagination like the force created by the two sailors discussed above. John Archibald Wheeler, the physicist who is known for having coined the term black hole, backs me on this for he said, "Should we be prepared to see some day a new structure for the foundations of physics that does away with time? . . . Yes, because "time" is in trouble. These lines about John Wheeler were written in Paul Davies' book, About Time, page 178. I saw John Wheeler a couple of times in the Pepin Physics lab when I was attending Columbia University. He always wore a khaki trench coat. He was responsible for reviving interest in general relativity in the United States. He worked with Niels Bohr in the field of nuclear fission. He was a collaborator of Albert Einstein. Sadly, he died before achieving Einstein's vision of a unified field theory, which still remains not unified.

CHAPTER 16

Did We Invent Time the Same Way We Invented Money?

Usually things that exist in a physical world will interact with that world. If time exists in this world it should interact with this world. If it does interact, it does it in a different way compared to other physical entities, such as matter. One thing, I think, physicists agree upon, is that time reversal does not affect any physical process. In all the laws of physics that we have found so far, there doesn't seem to be any distinction between the past and the future in regards to the flow and the path of time. The laws of physics still hold even if a physical process goes backward in time. There is a one-to-one relationship between a process and its time reversal. For every big bang there is a big crunch; you just have to run the movie backward. If a dish falls down from a table, it can go up to the table; you just have to give back the energy lost to the act of falling, and the physical laws will allow the dish to go back to the table and glue its pieces up and be a dish again. Some might say that the increase of entropy will not allow you to go back because of the irreversibility of the process. Irreversibility occurs when you have a loss of energy. If you give back that energy loss the process can go backward. According to Heisenberg, Schrödinger, and Boltzmann, any process can go backward in time as long as you give back to the system the energy lost. The only catch, here, is in some processes it will take a long time.

You should remember that physical laws allowed people to exist, even though they are more complex than a dish. If you take away some of the energy from the dish-table system, the physical laws or Nature will not allow the dish to fly up to the table and sit on its surface the same way it was sitting before the act of falling.

Physicists assigned this time reversal thing a letter called T. It is like an operator. It will act on a process and make it run backward. In physics there are many operators.

Was time invented out of necessity? Money didn't exist before people existed. When there were only dinosaurs roaming our earth, there was no money. People were the creatures that invented money. People invented money to buy things from each other. First, money was probably made of seashells, and then as we progressed it turned into metal and paper and plastic and so on. Now, the American dollar, for example, is worth only the trust of the United States Government. There is probably no gold stored in the United States Government treasury to match the number of American dollars circulating in the world. The American dollar is protected by the trust of the United States Government only. If the United States government collapses the dollar will collapse with it. There is not enough gold to protect the dollar. And this is true, whenever there is some instability in the American Government the Dollar goes down in value and the price of gold goes up. In facts, the price of gold goes up whenever there is some major instability in the world.

Money was invented because people needed it. Now money is becoming a major part of our lives. People kill for money. A survey in England was made asking people if they could kill a human being for one million dollars. Many people said yes with the condition that they don't get caught. People who sell their body for money are prostitutes. Some people sell their soul for money; they marry somebody only for money. People kill their parents for money. Governments kill thousands of people because of money. Money won't go away. It will only go away after every person on Earth dies. So money is stuck with us. It has become like cancer. It takes on many forms so that we can't get rid of it. If we only use credit cards, money will become credit, and credit will play the role of money, and we will get stuck with credit. We cannot escape from money now. We are trapped by money and it has become the mechanism that runs our lives. I understand that it is not money that counts but the thing that money buys. I also understand that it is not time that counts but the thing you do with your time.

I thought that time was like money. People needed to meet to conduct business or go hunt together. They needed something to help them meet before they would go for the hunt, especially if what they were hunting for was a large and dangerous animal. They needed to gang up against that animal, so they needed to go as a group into the woods together, or to the savanna or to whatever hunting place there was at the time.

They would probably dig a long stick into the ground and say to their hunting partners when the shade of the stick gets to this point or to that point on the ground we will meet. It could also be that they pointed to the sun and said when the sun is in that location between the zenith and the east horizon we will meet. Or if the sun is right there above our heads, we will meet and eat together. Our ancestors, for sure, used the sun as a clock.

Now time has become a part of us. Some people die because of time—the emergency room was so crowded and the nurses couldn't tend to him, so he died. Some people get killed because of time—If she doesn't bring the money on time, she will get shot by her pimp. Some people live because of time—the governor gave him more time to investigate his case and they found him innocent and his life was spared. Time now is imbedded in our lives. We cannot live without time, the same way we cannot live without money. Of course you can live without money if you find yourself a place with animals, vegetation, and water. Although, it will be hard to find a place like that without a government coming, asking you to pay taxes for it with money.

Now we need to ask ourselves where this time came from. Like I stated earlier, I once met Murray Gell-Mann, the father of the quark model. When I had asked him "where did the Big Bang start?", he answered me by saying, you cannot ask that question because there was no space before the Big Bang, which means there was no where. He also said that there was no time before the big bang, so there was no when. In his case, time was created at the beginning of the big bang. So now, according to the majority of physicists, we know when time was created. It was created with the big bang.

Sidi at Columbia University Physics Department

CHAPTER 17

Explanation of my Appendix paper

In 2001 I was a visiting assistant professor at Wittenberg University, a private institution in Springfield, Ohio. I wasn't happy teaching there. The city, Springfield, was gloomy, and there was nothing going on that would make you feel at home. There were more gas stations than coffee shops. And the coffee they serve at those gas stations was always diluted to the point that you could see the bottom of the paper cup. There was some poverty too. I came from Minnesota. Minnesota seemed rich compared to this location in Ohio. There were a lot of white people who looked poor, a thing I didn't encounter in Minnesota. They had bad teeth and depressed faces, probably from extensive labor or heavy drinking. It seemed to me like nothing exciting was going on. Of course, there were few people who knew where one could have a good drink and socialize. I was new to Springfield and all I knew about was this cool, small adjacent city called Yellow Springs. It was about a 15 minute drive from Springfield. I visited this small cool town several times to the point that I became a regular at this coffee shop.

There was a small college, still is, near the shop called, Antioch College. It had only a few students in its registry, and was an underdog, so to speak, compared to Wittenberg University. It had been active and vibrant during the Vietnam War, but now it was just a bunch of old empty buildings. The library leaked when it rained and its students, if you could find them on campus, looked different. They looked like hippies. I liked Antioch College more than I liked Wittenberg University. Antioch College was real to me when Wittenberg University was phony. I used to go to Antioch's College main library and write my time and relativity paper. Beside a couple of students, I was the only person using the library. You would sometimes find a couple of students typing on the computer and that was about it. The school was dying even though it had a great history. I would walk into the library, pick a table, sit down and start to write. In my paper, I added an imaginary term to time, T. I wanted to know what would happen to that extra term when I insert this time, T, in the equations of Einstein special theory of relativity. What happened was at the end, I ended with two terms, A and B, when multiplied by each other, resulted in a product of zero. The second term, B, was the imaginary part of time, which I added in T. The A part was the one that contains Einstein's equation of special theory of relativity. The product A*B=0. If A is equal to zero, we get Einstein's equation, and if B is equal to zero, the time will lose the imaginary part I added in the beginning. If the product is equal to zero and we know that A=0 gives us Einstein's equation, that means that B can be zero or not zero, which means that time can have an imaginary part. I was happy when everything fit. I checked my calculations more than once, and whenever I checked it, it turned out to be mathematically correct. I said to myself, before I do anything, I need to send it to some physicist I know and see what they would say. I sent it to my friend, Bayram Tekin. Bayram Tekin was a very smart guy. He is also very blunt. If you make a mistake in some calculation he would call it stupid. I remember once I was reading a physics paper and I didn't understand it. I went up to him and I asked what should I do with this paper, I don't understand it? He said, you should rape it, not just have sex with it. I was surprised he said that because he was a conservative Muslim. Anyway, I sent it to him as an attachment. He was a postdoc at Brandeis University at the time. I was waiting for his email expecting a response that would say, "this is stupid, Sidi. What did you do?" He responded three days later, saying that space picked up an imaginary part. I didn't see it in my paper, but he saw it. I didn't like the idea of space picking up an imaginary time because I wasn't as much interested in space as I was with time. Now why would space pick up an imaginary part if time picked up an imaginary part? Does it mean that they are related, or tied together? If they are related how are they related? Physicists claim that time is the fourth dimension. If the fourth dimension here is related to space, does it mean that one dimension of space, such as x, is related to another dimension, such as y? Mathematicians claim that a vector pointing in the x direction has to be independent of another vector pointing in the y direction. We know that this independence is mathematical. Does it have significance in the physical world? I didn't stop here. I sent my paper to my advisor, professor Benjamin Bayman. I have so much respect for him. If it weren't for professor Bayman, I wouldn't have gotten my PhD. He helped me with my thesis and showed me a great deal of physics. I sent him my paper and waited for an answer, knowing exactly that he would respond. I wasn't afraid because I knew my paper passed Tekin, my intimidating friend.

After a couple of days he responded: "if this is true, then it is easy to show that two 4-vectors have an invariant scalar product, if the 4-vectors x and y are both subject to the same Lorentz transformation. The derivation of Equation (4) from Equations (1,2, and 3) is purely algebraic. No assumptions have to be made about the reality of the x and y. For this reason, I don't find it surprising that you can subject a 4-vector with an imaginary time component to a Lorentz transformation. It seems to me that if you want to pursue this subject, you have to demonstrate the physical significance, or usefulness, of having an imaginary time component. Until you do this, I don't think you have gone beyond a re-statement of the familiar equations given above.

Best wishes."

CHAPTER 18

Schrödinger Equation and Time

Schrödinger equation is an equation, that when solved, gives information about the energy or the behavior of the particle you are studying. In facts, there are two equations: Time-dependent and time-independent Schrödinger equation. I took the time-dependent equation and I added an imaginary part of time to see what would happen to the imaginary part once I solved for the energy. The energy is a real, physical entity. I said to myself, if energy ends up with an imaginary part, time has influenced the energy. I also wanted to check the influence of the imaginary time over the behavior of the particle I am studying. This behavior shows up in what we call the wave function. If the energy turns out to be real, regardless of the imaginary part of time, then the time-dependent Schrödinger's equation is not sensitive to time or doesn't really care about time, so to speak.

Sofie Leon was an undergraduate student at California Polytechnic State University in San Luis Obispo and she assisted me in this research. We added an imaginary part to time and inserted this new time to the time-dependent Schrödinger equation to see if the added imaginary part will affect the solution. I wrote a relativity paper discussing what we had found: the extra term did not affect the solution and doesn't have to be zero to get the exact solution of the equation. Now we need to ask ourselves which case is worse; Doing physics without time while time really does exist as a physical property, or doing physics with time while time doesn't really exist as a physical property? One can do physics in a two-dimensional space without violating the laws of physics. You will get a different picture of a physical process, but that physical process will not violate the laws of physics. For example, if you have a system that is moving in spiral in the z-axis direction, by doing your calculation in two-dimensional space you will think that the object is moving in a circle. Your calculation will not violate the law of physics, but you added another dimension that doesn't exist physically so the effect of this dimension will show in the result and the result might be wrong or unphysical because it contains an unphysical dimension. It is possible to solve a problem without the use of two spatial dimensions when it is moving in only one direction. For example, we can study the motion of a car on the highway. Even if we are in a three-dimensional space, we can still solve the problem of the motion of a car without the need of the other two spatial dimensions and no harm is done. Now, if we add an unphysical dimension to our analogy, we will encounter a problem. Reading this and understanding it can give us an idea that doing physics with this addition of time can give wrong results in the study of physical processes that are sensitive to non-physical things. When a light of some frequency travels from one medium to another, its frequency doesn't change. Why is that? When a light travels from one medium to another, does its energy change? $E=h.f$. If its frequency doesn't change, according to the formula, its energy doesn't change either. Is that true?

When light passes through a medium, its energy doesn't change, unless it is absorbed. The reason the light slows down is because it interacts with the particles of the medium. The atoms absorb light and then give it back. Because of this process, the light slows down. The medium might get warm, which means that the medium is absorbing energy. The photons that get absorbed excite the atom, and this excitement shows in a form of heat. So, some photons make it through and other gets absorbed. The energy of the photon doesn't change, which means that its frequency doesn't change. The frequency is the inverse of the period and the period is time. So the time of oscillation doesn't change. Now, we need to ask ourselves, why doesn't the frequency, or the period of one oscillation, change? Is it possible that it doesn't change because it is not a physical thing? Since the question of time is at the center of special relativity, one cannot investigate time without investigating special relativity. Is there a connection between energy and time? If time is connected to space and space is connected to energy, then we have a connection between time and energy. We know that space is related to energy in a way to manifest itself. You need a room for energy to propagate or dissipate. One cannot imagine energy without having a space to manifest itself in. Scientists have related space to time in a loose connection. Time is an extra dimension; one that has to be added to space. This connection is somewhat loose because time was just added by hand, so to speak. Time was not extracted from space. We know that space is tied with time through the formula $x=ct$, where x is some distance which belongs to space, c is the speed of light and t is time. If, for example, we know the age of the universe, we can calculate the size of the universe since the particles that are now reaching far out in the void are the massless particles, and these massless particles can only travel with the speed of light. The uncertainty principle states that energy and time are tied together in the equation, $Et=h$. Since energy is quantized, can we declare that time is also quantized? But what is the use of a quantized time? One can state that if we make use of quantized energy, we can probably also make use of quantized time. If there is a connection between energy and time, this would prove that time exists. Simply put, if one exists, the other also has to exist. A mother who gives birth to a child physically exists and is real therefore, the child also exists and is real. Did space give birth to time? Did energy give birth to time? We know from physics that time didn't come out of space; it is just another dimension that was added to space. Heisenberg's uncertainty principle is a loose connection since it doesn't indicate where time has come from. John Wheeler said, "Should we be prepared to see some day a new structure for the foundations of physics that does away with time? . . . Yes, because 'time' is in trouble." Check Page 178 of the book titles: About Time, written by Paul Davies.

The subject of time in quantum physics is a bit cloudy, which is ironic because quantum physics is supposed to oversee almost all kinds of physics including gravity. You see this quantum fuzziness of time in Heisenberg uncertainty principle. Now that we have time in quantum gravity, any uncertainties or complications that are related to time will manifest itself in quantum gravity.

CHAPTER 19

Kurt Gödel and Time

In 1949, Kurt Friedrich Gödel had produced a remarkable proof which states that in any universe described by Einstein Theory of Relativity, time as we know of cannot exist. He showed his work to Einstein. Gödel and Einstein became good friends while working together at the Princeton Advanced Institute. In fact, it was Einstein who helped Gödel obtain his American Citizenship. Gödel was a German citizen and the US government, back then, was suspicious of foreign scientists, Germans in particular. Einstein endorsed Gödel's proof but to a certain degree, since it determinedly overthrew his new classical idea of the world. Einstein could find no way to disapprove it, and in the half-century since then, neither has anyone else.

What's so interesting is that when Kurt Gödel was preparing for his American citizenship exam, he found a flaw in the U.S. constitution. This flaw would allow the U.S. to become a dictatorship. Einstein advised Gödel not to mention that during the interrogation because it could jeopardize his chance of getting the citizenship.

Can one come up with an experiment to show that time is a physical variable like space? What experiment should a scientist come up with? Before we ask this question about time, let's look at a simple case. For instance, like doing an experiment to show that a particle exists. Theoreticians sometimes would suggest or predict the existence of a particle on paper and the experimentalists would run to the lab and search for it in the real world and find it—this occurs often. Murray Gell-Mann, the father of the quark model, had projected from his model that a particle made of three strange quarks, now called the omega minus, should exist. Experimentalist ran their experiments and found that the particle truly does exist in nature. Sometimes, experimental physicists would look for particle A only to find particle B instead. To find these particles, physicist had to generate collisions. Since these particles are made of mass or energy, they would interact. This is easily done with particles, but can it be done with time? If time is made of a physical entity, we don't know what this physical entity is. Particles have an effect on other particles. Does time have an effect on another time? Does time have an effect on space? Does time have an effect on particles? Some might say that time has an effect on the neutron, since neutrons decay within fifteen minutes. If time affects the neutron decay, how come it doesn't affect the electron? The electron doesn't decay. Would the neutron decay if time didn't exist? We know that muon decay is affected by relativity. Does time have an effect on physics? We know that an electric field affects the motion of an electron. We know that the magnetic field affects the motion of an electron. We also know that the gravitational field affects the motion of an electron. Does time affect the motion? As a physicist, we know that time doesn't affect the motion of an electron. If it doesn't affect it that means that there is no interaction between the two. If there is no interaction, it doesn't mean that one or the other doesn't exist. Two photons can't interact with each other even though they both exist. The photon consists of oscillating magnetic and electric fields but that magnetic field doesn't interact with the magnetic field of another photon, why is that? To answer this question, I conducted an experiment with my student, Maria Gilbert. We generated some fringes from a laser beam and a double slit. We sent the laser beam through the gap of a large magnet with the magnet turned off. We looked at the fringes against the wall and we turned on the magnet to see if the magnetic field of the magnet will interact with the photon or its oscillating magnetic field. We didn't see any changes.

CHAPTER 20

Questions about the brain

There are over 75 organs in the human body. Our organs have many vital functions, but are limited in their operations. Our eye allows for vision, but it can only see so far before things appear unclear. For example, we can't see a flower on a mountain three miles away. Our ears allow for hearing, but can't hear faint noises, such as the noise of an ant walking. Our nose allows for smell, but can't smell weak odors such as the scent of a rose five miles away. Maybe the reason our organs are not that efficient is because we don't need an extreme efficiency of our organs to survive in this world. Our brain is also an organ. These organs are all tied to our head and they all need oxygen from our blood as well as nutrients. Can we conclude from this that our brain is also inefficient? If our brain is inefficient, this means it cannot think of certain things that are outside the realm of our survival as species. Our eyes, ears, and nose are made to the extent for us to survive as species. So, we can state that our brain is also made to the extent for us to survive as species. Now, do we need to know where we came from for us to survive? Do we need to know who god is or what was before the big bang for us to survive as species? Animals survive without knowing where they came from. Bugs survive without knowing where they came from. I once had an interview for a physics teaching position at the Claremont colleges in Claremont, California. The interviewer was one of the physics professors. He asked his students to have lunch with me so they could get an understanding of who I was and how I behaved towards them. During lunch, I asked the students "Do we need to know where we came from for us to survive as species. In other words, Does our brain have to be structured in a way to know where we came from, or who is god, if there is one, for us to survive as species?" Some students didn't know what to say, but one female student said yes, we do need to know. However, when I asked her why, she didn't give me an answer. Nature is conservative and it can't give you more than you need. For example, when an object falls from a building, the energy it has at the bottom is the same or less than the energy it had at the top before it started falling. It can't have more at the bottom. Nature will not give more energy than the laws of energy conservation allows. My point here is that nature makes things in a way to minimize energy. Our brain is made of particles, which are of nature, so we can conclude that its structure is made in a way to minimize on the energy. The structure of our brain is probably made under some constraints which minimize the energy. Nature will not allow us to be more intelligent than we are supposed to be to survive as species. It seems like the most important thing in this world is life. There is a force of life that pushes us to procreate. This force of life is exerted on animals too. So, it is possible that our brain is limited and questions about the origin of life and origin of the universe cannot be answered because our brain is structured in a way to find only the answers of the questions that are related to our survival in this world. Almost everything that we make now is important for survival. For example, we manufacture cars and planes for transportation of people and goods. We manufacture TV's and phones, which are necessary for communication. It seems like everything we make is tied to our survival without us thinking about it. We produce things automatically. If you invent something that is not useful in our survival, you won't get the fund for it to produce more of it. You only get funded from banks or investors if they know that your invention is practical and going to sell. So we think only to survive. We can't think of unnecessary thoughts. The question becomes, why do we ask the question? Is asking the question a proof for us to have to find the answer? Is asking about the physicality of time critical to our survival as species? It seems to me that our intelligence is beyond what we need to survive because there many animals that have been living on this planet and they have less intelligence than we do.

Why is that we have extra intelligence? Can this extra "useless" intelligence that we have help us understand the extra things, such as the physicality of time? It is possible that the physicality of time is unnecessary for our survival.

Another problem of time, like I stated before, is that it doesn't distinguish between past and future. Feynman said, "In all the laws of physics that we found so far, there does not seem to be any distinction between the past and the future. The law of gravitation is of such a kind that the direction does not make any difference; if you show any phenomenon involving only gravitation running backwards on a film, it will look perfect. It is easy to prove that the law of gravitation is time-reversible. We believe that most of the ordinary phenomena in the world, which are produced by atomic motions, are according to laws which can be completely reversed. The laws of molecular collision are reversible." If you go up to the atmosphere from the surface of earth, you can figure out how high you are from the surface by knowing the density of air. So, you can distinguish between up and down, but with respect to time, you cannot distinguish between past and future. Of course, in the large scale realm, if you look at the processes that occur in our daily lives, you would know that time passes when your dog gets old or the tree grows tall. But in the small scale, like in the particle physics scale, we won't be able to distinguish between the two. A particle penetrating a field can be looked at as a particle leaving the field. In its past, it could be that it has just left the field and is now moving away from it; but another person might state that in its past it was away from the field and now it is penetrating it. So you don't know the past from the future of this process. The time variable, physicist work with, fails to detain the idea of time as we live it. As physicists continue to try to devise more physical laws of nature, the time variable disappears altogether.

CHAPTER 21

Plants, animals, and Time:

What are the things that exist in this universe? We know that matter exists. One can state that everything around us is made of matter: People, the sun, the planets, the galaxies, etc. Energy comes from matter and vice versa, so energy is a form of matter. The universe is made of space and matter. Some philosophers might debate the existence of space. But as a physicist, I don't have to worry if I confirm that space is non-negotiable in regards to its existence. One confirming reason is we need space to place things in. We need space to build a house or to place your old car that you don't want to get rid of. We need matter and space to survive. Water is made of matter; we need water, therefore we need matter. We need space to survive. We need space to live in. Now do we need time to survive? This might sound like a hard question, but it is not. We don't need time to survive. Animals and plants need space and matter to survive, but they don't need time. They don't have watches and they don't need them. They go by feelings. When it starts to get cold, they will migrate south. Animals don't care about the day or the month; they only care about how it feels during that time. Plants do the same thing. If the temperature changes to cold and stays cold for a while, the plants will start to change physically. The color of the leaves might change. Some plants will start to bud as the temperature begins to rise, regardless of the day of the month. There are events or processes that occur in a plant in order for it to survive or grow. Plants and animals don't need time to survive. We can say the same thing about us; we don't need time to survive as species. Time is good for scheduling; it is a tool for measuring events, the same way the ruler is a tool for measuring the length of a table or the length of a room. Just because we don't need something to help us survive, doesn't mean that the thing doesn't exist.

CHAPTER 22

What is Reality?

According to a typical dictionary, reality is the fact, state, or quality of being real or genuine, that which is real; an actual thing, situation, or event. For example, when a person dies he or she has experienced the reality of death. Death is a reality in the same sense that life is a reality. When a child is born it is real. Personally, I believe reality is something we do not know exactly like we think we know. What the dictionary says is what humans define. It is possible that our reality is not the true reality. Reality is a perception, and we only perceive what we experience in life, but there are infinite things to experience, and we cannot experience all of them because we die. Our lifespan is short. When we look at the stars in the sky, to us, we feel like they live forever. When stars "look" down at us, however, they see us as flashes of light that live for a millisecond—their time—and then die. To stars, we live only a very short time. Since we only live a short time, we are only able to leave a little bit of what we experienced or what we learned for the coming generations. Like I stated before, we don't live long enough to watch the whole game of life and fully understand its rules. Imagine someone is the only person in the world who knows the rules of the game of chess and he is teaching it to his daughter and when he is about to finish explaining to her all the rules of the game he has a heart attack and dies. His daughter now doesn't know all the rules. So when she tries to teach her son how to play chess, she will not be able to explain to him all the rules. This lack of knowledge will propagate through out the descendents of this family and this family will never be able to know all the rules of the game of chess. The sad thing is people will start making new rules to try to complete the game. And these new rules will change the game. This is exactly what's happening to us: we make new rules. And we do so because we don't know the true rules of life. We don't know the true meaning of life. We make our own reality. Humans usually make a reality that pleases them, a reality that will make them feel good. Like the old, wrong reality that states that all the planets in our solar system revolve around the Earth and the Earth is at the center. We were making the Earth the center of the solar system or the Universe because we thought we were important and consequently our Earth has to be

A Quest for the Physical Reality of Time

as important, when in facts, Earth is just another planet revolving around the sun like the other planets. We created a reality based on how we perceive ourselves.

When you ask "what is reality?", it is like asking "what is the truth?". If you are religious you might be easily convinced about the truth, but if you are not religious and you want to know the truth you will encounter a difficult task. When you ask "what is reality?" you are automatically giving an existence to it. How do you know that reality really exists? As I've said, if a scientist wants to prove that something is physical, he or she needs to make an experiment or write a theory that will be tested by an experiment. To prove that a particle exists, for example, a scientist needs to first calculate the probability for such a thing to exist. Unfortunately, everything has a certain probability for existing. For example, in quantum mechanics, there is probability for a particle to tunnel through a high potential wall even though its energy is less than the energy of the potential wall that is blocking it. The particle is not supposed to go through that wall, but eventually it will do, because there is a probability that it will go through. This phenomenon is called tunneling.

When I was an undergraduate student, one thing in quantum mechanics that bothered me a lot was the probability of finding a particle in a portion of a volume. In quantum mechanics a particle is governed by a wave function. It is a mathematical function like $f(x)=x+2$, for example. In knowing the function, you have some information about the particle. When physicists want to calculate the probability of finding a certain particle, say an electron, in a small volume, they get the wave function, which, in our case is f(x)=x+2. Then it is multiplied by its complex conjugate, which is f(x)=x+2, since f(x) is real here. Finally, all probability densities are summed up throughout the whole volume of interest. Like I have told you before, I received help in understanding this idea from T.D. Lee. He explained to me that physics is this type of inductive reasoning. After some thought, I understood what he was talking about. This type of inductive reasoning is how we conclude that copper is conductive.

The problem with physics is there is nobody who would say that energy conserves, parity conserves, and momentum conserves. We simply assume that they conserve until somebody does an experiment and tells us otherwise. This is exactly what happened with the issue of parity. Physicists used to think that it conserves until, Lee, Yang and Chu proved that it doesn't. Now how is this related to the subject of this book? Well parity is an operator and Time can also be an operator, as I discussed previously. So if we proved that parity is violated that means that it probably exists in the physical sense. Thus, if we prove that the Time reversal operator is violated, I believe that it will show that time exists in the physical sense—this believe is just a random shot in the dark. Of course, this is just a thought. In quantum mechanics we only calculate probabilities. If I want to know if an interaction existed between a gluon and a quark, or an electron and a photon I would need to calculate the probability for such interaction to occur. This made Einstein mad and pushed him into saying that God doesn't play dice with the world, which means that God had created the Universe—That is, if you believe that God created the universe. Remember that I am writing this book to all kinds of people, the religious, the atheist, the construction worker, the poet, the housewife, the restless teenager, etc.—without calculating probabilities; God already knew where things were. Quantum mechanics does not tell where a particle is, it only tells a region where it might be. So, we might state that if we want to know if time exists we need to calculate the probability of its existence. And even if we find some probability for it to exist, it will be like calculating the probability for it not to exist. If we take the number 1 and subtract from it the probability that time exist, we will find the probability for it not to exist. Finding the probability is good, but not good enough. By calculating the probability, we can determine if the particle exists. To do that, we need to know about the particle and its behavior. This means we need to know the wave function of the particle and then from that wave function, we calculate the probability density. Now you ask yourself what is this probability density? Let me describe it in a somewhat skewed explanation. If we understand the number density, we can understand the probability density. The number density is the number of things you are interested in measuring or knowing about divided by the volume where these things are. If you assume you have 10 books in a room which is 10 meters high, 10 meters long, and 10 meters wide, the number density of books is 10/(10X10x10), which is 1/100 books per meter squared. So it's just the number of books divided by the volume. We can use the same example for the probability density, the only thing we need to do is replace books with probability. Now this probability density is somewhat confusing when you start talking about particles and the wave function. If there is such density, the particle exists and is located in the place or near the place where we thought it would be. Now, we need to know the wave function. A wave function is a mathematical function where the information of that particle is stored, such as the energy, the mass, etc. Sometimes we put some real conditions to make the wave function more realistic. One condition is to normalize the wave function, which means that if we sum all these probabilities densities, through the whole universe, we should get 1, which means that the particle has to exist somewhere. If it doesn't exist here it should exist there. Normalizing the wave function makes the wave function more manageable. Now, can we normalize the wave function of time to calculate the probability of the existence of time? If you normalize it this will mean that time has to exist somewhere in the universe. That's a big thing to assume. If you can't state that, it becomes very hard to calculate the probability of the existence of time. Einstein found the probability approach to be unsatisfactory in finding particles, let alone time. Einstein was unhappy with quantum mechanics because it only calculates probabilities. Einstein didn't like probabilities because he thought God didn't use probabilities in making the universe. He felt it was a short cut, or a lazy way to describe nature. He wanted to describe nature in a way that was very concrete and not based on probabilities. He stated that God doesn't play dice. When God was making the world he didn't build it based on probabilities; he built it based on facts. If God didn't play dice, then he did not want to either. He did not want to base his findings on probabilities. Einstein felt that God didn't like short cuts. People use short cuts to economize on time, space or whatever, but God has plenty of everything: plenty of time, plenty of space, and plenty of energy. What was so confusing about Einstein were his beliefs about God. Once, somebody asked Einstein who God was and Einstein said his God is the God of Spinoza. Spinoza is a philosopher who said that the book of religion is necessary for salvation among people who do not possess high intellectual gifts, namely the majority of people, but it should in no way limit the free exercise of the intellect in the search for the truth. Poor Einstein died without knowing the truth, without knowing God's facts. People admire Einstein in so many ways. They admire him because they admire themselves. They conclude that if Einstein could achieve such a high level of intellectuality, then all humans have the potential to reach that level. Einstein was like their extension to the unknown, like a finger that poked the belly of God.

APPENDIX

An Investigation of Time in Relativity

Sidi Cherkawi Benzahra[1]

Wittenberg University

Springfield, Ohio 45501

Abstract

I added an imaginary term by hand to the proper time, ΔT, of special theory of relativity and wanted to see what will happen to it if I tried to get the time dilation relationship between $\Delta T'$ and ΔT. I found that this added imaginary term of time does not have to be zero.

If ΔT is the real, proper time of travel of a light beam from one point to the other,

$$\Delta t = \Delta T + i\Delta \xi \qquad (1)$$

Δt will be the time I choose and work with in order to see what will happen to the imaginary part $\Delta \xi$ once I apply Einstein's special relativity. $i\Delta \xi$ here is an imaginary term which I added by hand. Consider two observers O and O' [1]. O fires a beam of light at a mirror a distance L away and measures the time interval 2 ΔT for the beam to be reflected from the mirror and to return to O. Observer O' is moving at a constant velocity u. As seen from the point of view of O, the beam is sent and received from the same point, while O' moves off in a perpendicular direction. The beam is sent from some point A and received at some point B a time 2 $\Delta T'$ later, according to O'. The distance AB is just $2u\Delta T'$. According to O, the light beam travels a distance $2L$ in a time $2\Delta T$. According to O', the light beam travels a distance of $2\sqrt{L^2 + (u\Delta T')^2}$ in a time $2\Delta T'$. According to Galilean relativity, $\Delta T = \Delta T'$, and O measures a speed c while O' measures a speed $\sqrt{c^2 + u^2}$. According to Einstein's second postulate, this is not possible–both O and O' must measure the speed c. Therefore ΔT and $\Delta T'$ must be different. I can find a relationship between ΔT and $\Delta T'$ by setting the two speeds equal to c. According

[1]sbenzahra@wittenberg.edu or benzahra@physics.umn.edu

to O, $c = 2L/2\Delta T$, so $L = c\Delta T$. According to O', $c = 2\sqrt{L^2 + (u\Delta T')^2}/2\Delta T'$, so $c\Delta T' = \sqrt{L^2 + (u\Delta T')^2}$. Combining these, I find

$$c\Delta T' = \sqrt{(c\Delta T)^2 + (u\Delta T')^2} \tag{2}$$

and, solving for $\Delta T'$,

$$\Delta T' = \frac{\Delta T}{\sqrt{1 - u^2/c^2}}. \tag{3}$$

This relationship summarizes the effect known as time dilation. But in my calculation, where $\Delta t = \Delta T + i\Delta \xi$, I have according to O,

$$c = \frac{L}{\Delta T + i\Delta \xi}. \tag{4}$$

I can see here in the equation just above that since I added an imaginary part to time, length in particular or space in general picked up an imaginary part also. According to O'

$$c = \frac{\sqrt{L^2 + [u(\Delta T' + i\Delta \xi')]^2}}{\Delta T' + i\Delta \xi'}. \tag{5}$$

Taking the expression of L from equation (4) and inserting it in equation (5) and squaring, I get

$$c^2(\Delta T' + i\Delta \xi')^2 = c^2(\Delta T + i\Delta \xi)^2 + u^2(\Delta T' + i\Delta \xi')^2. \tag{6}$$

Separating the real terms from the imaginary terms of equation (6) I get

$$[\Delta T'^2(1-u^2/c^2) - \Delta T^2] - [\Delta \xi'^2(1-u^2/c^2) - \Delta \xi^2] + 2i[\Delta T'\Delta \xi'(1-u^2/c^2) - \Delta T\Delta \xi] = 0$$

I can see that time dilation relationship is given by setting the first part of the equation above equal to zero

$$\Delta T' = \frac{\Delta T}{\sqrt{1 - u^2/c^2}}. \tag{7}$$

I can also see that there is an effect of time dilation in the imaginary part, $\Delta \xi$, given by

$$\Delta \xi' = \frac{\Delta \xi}{\sqrt{1 - u^2/c^2}}. \tag{8}$$

Since a complex number vanishes only if its real and imaginary parts vanish too, I get

$$\Delta T'\Delta \xi'(1 - u^2/c^2) - \Delta T\Delta \xi = 0. \tag{9}$$

Taking the right-hand side of equation (8) and inserting it in equation (9), I get

$$[\Delta T'\sqrt{1 - u^2/c^2} - \Delta T]\Delta \xi = 0. \tag{10}$$

Equation (10) is the product of two terms. Since Einstein's relativity is considered to be true, the first part vanishes, and leaves us with the condition that $\Delta \xi$ can be different from zero. And if $\Delta \xi$ is different from zero that makes Δt imaginary since Δt is equal to $\Delta T + i\Delta \xi$ as indicated in equation (1). I have mathematically found that relativity allows time to have an imaginary part.

References

[1] Kenneth Krane, Modern Physics (John Wiley and Sons, 1983).

www.ingramcontent.com/pod-product-compliance
Lightning Source LLC
Chambersburg PA
CBHW030755180526
45163CB00003B/1030